Aluminium Alloys - Synthesis, Properties, and Applications

Edited by Shashanka Rajendrachari and Adarsh Patil

Published in London, United Kingdom

Aluminium Alloys - Synthesis, Properties, and Applications
http://dx.doi.org/10.5772/intechopen.1008619
Edited by Shashanka Rajendrachari and Adarsh Patil

Contributors
Amlan Das, Chaojiang Li, Hansong Ji, Heng Liu, Hiroki Nakata, M. Muneeb Waheed, Madiha Rafaqat, Naveed Ahmed, Na Zhang, Osamah Abdulhameed, Shashanka Rajendrachari, Takeshi Nasu, Tatsuhiko Aizawa, Wasim Alshammary, Yuxin Yang

First published in London, United Kingdom, 2026 by IntechOpen
IntechOpen is the global imprint of INTECHOPEN LIMITED, registered in England and Wales, registration number: 11086078, 167-169 Great Portland Street, London, W1W 5PF, United Kingdom

For EU product safety concerns: IN TECH d.o.o., Prolaz Marije Krucifikse Kozulić 3, 51000 Rijeka, Croatia, info@intechopen.com or visit our website at intechopen.com.

British Library Cataloguing-in-Publication Data
A catalogue record for this book is available from the British Library

Aluminium Alloys - Synthesis, Properties, and Applications
Edited by Shashanka Rajendrachari and Adarsh Patil
p. cm.

This title is part of the Materials Science Book Series, Volume 27
Topic: Metals and Nonmetals
Series Editor: Chonghe Li
Topic Editor: Joan Josep Roa Rovira

Print ISBN 978-1-83635-707-0
Online ISBN 978-1-83635-706-3
eBook (PDF) ISBN 978-1-83635-708-7
ISSN 3049-8856

If disposing of this product, please recycle the paper responsibly.

IntechOpen Book Series

Materials Science

Volume 27

Aims and Scope of the Series

Materials science has always occupied an extremely high position in the human development process. As we explore the oceans of stars, various industries have put forward more stringent requirements for the performance of materials, forcing us to pay more and more attention to the development of new materials. At the same time, the formation of a data-driven scientific paradigm is dramatically shortening the development cycle of new materials. The huge data generated by synergistically combining theories, high-throughput experiments, high-throughput computation, and artificial intelligence is greatly contributing to our ability to utilize materials science to solve real-world problems. The three topics of this book series - Metals and Nonmetals; Composite Materials; and Surface Science - will address important areas of advancement in materials science. There will be a range of interesting works published under these topics.

Meet the Series Editor

Prof. Chonghe Li received his Ph.D. from the Chinese Academy of Sciences in 1995. From 1995 to 2000, he worked as a researcher at the Shanghai Institute of Metallurgy, Chinese Academy of Sciences, where he also served as director of the research laboratory. In 2000, he was appointed Professor at the Institute of High-Performance Computing in Singapore, where he worked on computation and simulation in materials science until 2004. Since then, he has been a professor at the School of Materials Science and Engineering, Shanghai University, China, as well as the director of the Shanghai Specialty Casting Engineering and Technology Research Center. Prof. Li's research focuses on titanium alloys, titanium-aluminum single crystals, intermetallic compounds, theoretical calculations, alloy design, and special refractory materials. His broad scientific expertise is well recognized by the scientific community around the world. He is a member of the editorial board of the journal Metals. As an author, he has published more than 200 peer-reviewed papers, 2 books, and over 40 patents.

Meet the Volume Editors

Dr. Shashanka Rajendrachari is currently a Professor in the Department of Chemistry at Global Academy of Technology, Bangalore, Karnataka, India. He earned a Ph.D. in the Department of Metallurgical and Materials Engineering at the National Institute of Technology Rourkela, India. He is the first recipient of the prestigious Powder Metallurgy Student of the Year award by the Powder Metallurgy Association of India for the year 2015 at IIT Bombay. His name is listed among the World's top 2% scientists for the years 2020 to 2024, as reported by Stanford University, USA, and published by Elsevier. He has published more than 85 research articles, one German patent, 22 books, and 20 book chapters. His research areas include powder metallurgy, nanomaterials, and electrochemical sensors.

Dr. Adarsh Patil is an Assistant Professor in the Department of Mechanical Engineering at KLE Technological University, Hubli, Karnataka, where he has been serving since 2012. He holds a Ph.D. in Mechanical Engineering from Visvesvaraya Technological University and an M. Tech. in Process Metallurgy from National Institute of Technology (NITK), Surathkal. His primary research interests include metal-matrix nanocomposites, additive manufacturing, the wear and mechanical behavior of materials, and composite reinforcements. Dr. Patil has published many international journal papers, conference publications, and book chapters. He has also contributed as a reviewer for several international journals and conferences. His funded projects span material characterization and the development of tribological facilities. In addition, he holds two patents in the fields of safety and agricultural technology.

Contents

Preface

Aluminum alloys play an important role in advanced engineering due to their low density, high strength-to-weight ratio, corrosion resistance, and extraordinary tensile strength. Because of these properties, Al alloys are widely used in aerospace, automotive, marine, electronics, and energy sectors. This edited book provides a brief, reliable overview of the fundamentals of aluminum alloys, including their principles, processing methods, surface properties, and corrosion behavior. These book chapters help readers understand fundamental concepts, advanced techniques, surface modifications, corrosion mechanisms, the evolution of microstructure in additive manufacturing, and more.

The book opens with *Introductory Chapter: Introduction to Aluminum Alloys*, which outlines the various aluminum alloy grades, their key properties, strengthening techniques, and applications in advanced engineering. Chapter 2, *Conventional and Energy-Assisted Milling of Aluminium Alloys* compares conventional machining with advanced techniques that improve machinability, tool life, surface properties, and process efficiency. Section 2 also includes Chapter 3, *Microstructural Evolution Mechanism in Fabrication and Micromachining of Selective Laser Melted Al Alloys*, in which the authors discuss how the microstructure of aluminum alloys transforms during selective laser melting and how this transformation affects mechanical properties and micromachining performance. Chapter 4, *Plasma Nitriding of Al-Cu and Al-Cu-Mg Alloys at Various Holding Temperatures*, discusses how plasma-based surface treatments improve mechanical and tribological properties and explains how processing parameters affect surface properties. Finally, Chapter 5, *Corrosion Mechanisms and Protective Strategies in Aluminium Alloys* explains, in detail, the basics of corrosion, their mechanisms, and how to protect the alloys from corrosion, including environmental effects, coatings, etc. Together, the contributions offer a balanced combination of foundational knowledge and advanced technological insights, making this book a valuable resource for postgraduate students, researchers, academics, and practicing engineers.

The Academic Editor and Co-Editor extend sincere appreciation to all contributing authors, as well as the production and editorial teams, for their dedicated support throughout the preparation of this volume.

Dr. Shashanka Rajendrachari
Professor,
Department of Chemistry,
Global Academy of Technology,
Bangalore, Karnataka, India

Dr. Adarsh Patil
Assistant Professor,
Department of Mechanical Engineering,
KLE Technological University,
Hubballi, Karnataka, India

Section 1

Fundamentals and General Overview of Aluminium Alloys

Chapter 1

Introductory Chapter: Introduction to Aluminum Alloys

Shashanka Rajendrachari

1. Introduction

Aluminum and its alloys are among the most important engineering materials of the modern era due to their combination of low density, good strength-to-weight ratio, excellent formability, corrosion resistance, thermal and electrical conductivity, and high recyclability. These attributes make aluminum alloys pivotal in aerospace, automotive, transport, packaging, construction, and electrical industries [1–4].

2. History and development

Commercial aluminum production began in earnest in the late nineteenth century with the Hall–Héroult electrolytic process. Over the twentieth century, alloy development (addition of Cu, Mg, Si, Zn, Mn, Li, and others) and improved thermomechanical processing methods produced a family of alloys tailored for different applications – from soft, highly conductive 1xxx series alloys to high-strength, age-hardening 2xxx and 7xxx aerospace grades [1, 5].

3. Chemical composition and classification

Aluminum alloys are grouped by the major alloying element and designated by the four-digit AA system. Major wrought alloy series include the following:

- **1xxx**: commercially pure Al (≥99%) – excellent conductivity and corrosion resistance.
- **2xxx**: Al–Cu family – high strength, used in aerospace; lower corrosion resistance.
- **3xxx**: Al–Mn family – good formability, used in beverage cans and sheets.
- **4xxx**: Al–Si family – welding and brazing filler alloys, as well as some extrusions.
- **5xxx**: Al–Mg family – marine and corrosion-resistant structural applications.

- **6xxx**: Al–Mg–Si family – medium strength, excellent formability, and weldability (common in extrusions and body panels).
- **7xxx**: Al–Zn–Mg family – very high strength for aircraft structures [5, 6, 7].

Cast alloys are identified by three-digit or four-digit numbering and different alloy chemistries (notably Al–Si alloys for casting). The choice between wrought and cast alloys depends on formability, mechanical requirements, and processing economics [3, 8].

4. Physical and mechanical properties

Aluminum's low density (~2.70 g/cm^3) gives it a high specific strength compared with many structural metals. Mechanical behavior, however, depends strongly on alloy chemistry and thermomechanical history: Precipitate-forming heat-treatable alloys (2xxx, 6xxx, 7xxx) can achieve much higher yield strengths via age-hardening than non-heat-treatable alloys (1xxx, 3xxx, 5xxx) that rely on solid-solution strengthening and work hardening [1, 9].

5. Microstructure and strengthening mechanisms

Key strengthening mechanisms in aluminum alloys include the following:

- Solid-solution strengthening: Dissolved Mg, Si, Cu, and Zn.
- Precipitation (age) hardening: Where nanoscale intermetallic/solute clusters formed during aging impede dislocation motion, a dominant route for 2xxx, 6xxx, and 7xxx series. The physics of precipitate nucleation, growth, shearing vs. Orowan bypass, and their multiscale modeling is an active area of research.
- Strain hardening: It results from cold work and grain refinement (Hall–Petch).
- Particle and dispersoid strengthening: It arises from stable intermetallics or Al3Sc/oxides in specialty alloys [9–11].

6. Processing: Casting, wrought processing, and heat treatments

Processing strongly controls final properties:

- Casting processes (sand, die, low/high-pressure, squeeze) are widely used for Al–Si-based alloys; microstructure control (SDAS, porosity, bifilms) determines mechanical and fatigue performance.
- Wrought processing (rolling, extrusion, forging) with controlled solution heat treatment, quenching, and artificial or natural aging tailors strength and toughness.

- Thermomechanical processing (controlled deformation + heat treatment) is used to optimize microstructure (grain size, texture, dispersoid distribution) [3, 8, 12].

7. Corrosion, surface protection, and fatigue

Although aluminum develops a protective oxide that gives it inherent corrosion resistance, many structural alloys (particularly those containing copper or with coarse intermetallics) suffer localized corrosion (pitting, intergranular) and stress-corrosion cracking in aggressive environments. Surface treatments (anodizing, conversion coatings), alloy selection, and proper design are necessary to manage corrosion. Corrosion modeling and predictive tools have advanced recently to better forecast lifetime and maintenance needs. Fatigue behavior is sensitive to casting defects, surface condition, and residual stresses [7, 13, 14].

8. Joining and fabrication

Aluminum alloys pose welding challenges due to high thermal conductivity, oxide films, and susceptibility to hot cracking in certain alloys. Gas tungsten arc welding (GTAW/TIG), gas metal arc welding (GMAW/MIG), friction stir welding (FSW), and modern filler alloys are widely used; 5xxx and 6xxx series are generally more weldable than 2xxx and 7xxx series without specialized techniques. Forming of aluminum alloys benefits from warm forming and control of strain paths to avoid cracking [5, 7].

9. Environmental and sustainability aspects – recycling

Aluminum recycling is energetically favorable: Secondary (recycled) aluminum production typically consumes typically ~ 5% of the energy required for primary production from ore. Secondary alloys are widely used in cast applications and increasingly in wrought components as recycling technologies and remelting practices improve. Challenges remain in the segregation of scrap, removal of contaminants, and maintaining mechanical properties after multiple re-melts; recent advances in preprocessing and remelting improve chip recycling and quality [15, 13, 16].

10. Applications and selection guidelines

Selection of an aluminum alloy balances strength, ductility, corrosion resistance, weldability, formability, fatigue behavior, and cost. Typical application examples are as follows:

- **Aerospace**: high-strength 2xxx and 7xxx series, Al–Li specialty alloys.

- **Automotive**: 6xxx extrusions, cast Al–Si components, and structural castings from recycled alloys.
- **Packaging**: soft 1xxx and 3xxx series for foil and cans.
- **Marine and architectural**: 5xxx series for corrosion resistance.

Designers also consider joining, surface finish, and lifecycle/recycling for sustainable choices [1, 5].

11. Current research directions and future trends

Major themes in contemporary aluminum research include the following:

- High-strength, high-toughness alloys achieved through alloy design and innovative heat treatments.
- Lightweighting via Al–Li and composite/laminate concepts.
- Improved casting integrity and additive manufacturing of Al alloys.
- Circular economy improvements, better scrap sorting, remelting, and alloying strategies for higher-quality secondary aluminum.
- Modeling across scales (from atomistic to component) to accelerate alloy design and predict behavior under realistic service conditions [9, 10, 16].

Aluminum alloys offer an unparalleled combination of properties for a wide array of engineering applications. Their development intertwines alloy design, microstructural control, advanced processing, surface engineering, and sustainability practices. Understanding the relationships between chemistry, processing, and performance is essential for the intelligent selection and innovation of aluminum-based materials.

12. Conclusion

The development of alloy design, from basic binary systems to sophisticated multicomponent and heat-treatable alloys, has made it possible for aluminum alloys to be used in extremely demanding applications, such as maritime constructions, automobiles, and aerospace. Their mechanical performance and long-term dependability are determined by the interaction of composition, processing, and microstructure. Aluminum recycling and green manufacturing techniques are becoming essential to future advancements due to the growing emphasis on sustainability on a global scale. The performance, sustainability, and efficiency of aluminum alloys in next-generation engineering systems are expected to be significantly improved by advancements in alloy design, additive manufacturing, and predictive modeling.

Conflict of Interest

The authors declare no conflicts of interest.

Author details

Shashanka Rajendrachari
Department of Chemistry, Global Academy of Technology, Bangalore, India

*Address all correspondence to: shashankaic@gmail.com

References

[1] Davis JR. Aluminum and Aluminum Alloys (ASM Specialty Handbook). Materials Park (OH): ASM International; 1993. 784 p. DOI: 10.1361/autb2001p351

[2] Altaş E, Rajendrachari S, Mahesh V. Introductory Chapter: Introduction to Aluminum Alloys. In: Rajendrachari S, editor. Recent Advancements in Aluminum Alloys. IntechOpen; 2024. p. 1–9. DOI: 10.5772/intechopen.113373

[3] ESAB. Aluminum Alloy Types & What They Mean [Internet]. 2024. https://esab.com/in/ind_en/esab-university/articles/understanding-the-aluminum-alloy-designation-system/ [Accessed: 2025-November-04]

[4] Trident M. Aluminum Alloy Designation [Internet]. 2024. https://www.trident-metals.com/wp-content/uploads/2015/01/Aluminum-Alloy-Designations.pdf [Accessed: 2025-November-04]

[5] Berlanga-Labari C, Biezma-Moraleda MV, Rivero PJ. Corrosion of Cast Aluminum Alloys: A Review. Metals. 2020;**10**(10):1384. DOI: 10.3390/met10101384

[6] Ruiz-Garcia A, Esquivel-Peña V, Godínez FA, Montoya R. Corrosion Modeling of Aluminum Alloys: A Brief Review. ChemElectroChem. 2024;**11**(9):e202300712. DOI: 10.1002/celc.202300712

[7] Nunes H, Emadinia O, Vieira MF, Reis A. Low- and High-Pressure Casting Aluminum Alloys: A Review. In: Rajendrachari S, editor. Recent Advancements in Aluminum Alloys. IntechOpen; 2024. p. 71–88. DOI: 10.5772/intechopen.109869

[8] Cui A, Wang X, Cui Y. Multiscale modelling of precipitation hardening: A review. Journal of Materials Science: Materials in Theory. 2024;**8**:13. DOI: 10.1186/s41313-024-00066-6

[9] Yusuf NK, Lajis MA, Ahmad A. Hot Press as a Sustainable Direct Recycling Technique of Aluminium: Mechanical Properties and Surface Integrity. Materials. 2017;**10**(8):902. DOI: 10.3390/ma10080902

[10] Zheng J, Pang Q, Hu Z, Sun Q. Recent Progress on Regulating Strategies for the Strengthening and Toughening of High-Strength Aluminum Alloys. Materials. 2022;**15**(13):4725. DOI: 10.3390/ma15134725

[11] ASM International. ASM Specialty Handbook: Aluminum and Aluminum Alloys. Materials Park. (OH): ASM International; 1993. https://www.asminternational.org/wp-content/uploads/files_main/pdf/ASM%20Subject%20Guide_Aluminum.pdf?srsltid=AfmBOorA_sOMmw1XxCzHiK1LOXviQc2ffLQ8ig9Y8KnyMiLKMRRHpI2t

[12] De Caro D, Tedesco MM, Pujante J, Bongiovanni A, Sbrega G, Baricco M, Rizzi P. Effect of Recycling on the Mechanical Properties of 6000 Series Aluminum-Alloy Sheet. Materials. 2023;**16**(20):6778. DOI: 10.3390/ma16206778

[13] Rubaye A, Abdulsahib H, Abdulwahid A. Corrosion Inhibition Properties of Norepinephrine Molecules on Mild Steel in Acidic Media. Journal of

Encapsulation and Adsorption Sciences. 2015;**5**(3):155–164. DOI: 10.4236/jeas.2015.53013

[14] Nunes H, Emadinia O, Soares R, Vieira MF, Reis A. Adding Value to Secondary Aluminum Casting Alloys: A Review on Trends and Achievements. Materials. 2023;**16**(3):895. DOI: 10.3390/ma16030895

[15] Chen X, Saada MB, Lavisse B, Ammar A. Recent advances in the remelting process for recycling aluminium alloy chips: A critical review. International Journal of Material Forming. 2025;**18**:42. DOI: 10.1007/s12289-025-01904-9

[16] Kok M. Production and mechanical properties of Al_2O_3 particle-reinforced 2024 aluminium alloy composites. Journal of Materials Processing Technology. 2005;**161**(3):381–387. DOI: 10.1016/j.jmatprotec.2004.07.068

Section 2

Forming and Machining of Aluminium Alloys

Chapter 2

Conventional and Energy-Assisted Milling of Aluminum Alloys

Naveed Ahmed, Osamah Abdulhameed, Madiha Rafaqat and Wasim Alshammary

Abstract

Aluminum, with its distinguished characteristics – that is, lightweight, durable, and highly recyclable – has become an excellent alternative to steel. Due to its growing industrial applications, the need for machining aluminum and its alloys has significantly increased. Aluminum and most of its alloys are considered relatively soft; therefore, the cutting forces and tool wear involved during milling are low compared to those associated with hard materials (e.g., super alloys). However, built-up edge, burr formation, material adhesion, continuous chip formation, thermal expansion, and residual stresses are inherent problems that occur during the milling of aluminum alloys, which adversely affect the tool as well as other performance measures, including cutting forces, cutting tool life, and surface quality. In this chapter, conventional milling and its challenges associated with the milling of aluminum alloys are discussed. Such challenges are addressed through the application of relatively modern techniques, such as external energy-assisted milling, which are also discussed in this chapter.

Keywords: aluminum alloys, chip adhesion, energy assistance, milling, tool wear and failure, surface quality

1. Introduction

In general, aluminum and its alloys are known to be easy-to-machine. In these alloys, the cutting forces are comparatively low, and the tool life is high, provided that there is no debris adhesion and the machining occurs without the formation of a built-up edge. The formation of continuous chips also results in low cutting forces and high tool life [1]. Different types of milling are used to machine aluminum and its alloys due to their increased demand in industrial applications as well as in daily life, such as furniture, window frames, and automobiles. Machining characteristics, or machinability, are determined based on several factors, for example, the tool deterioration metrics (wear, rupture, failure, and overall tool life), chip formation metrics (such as chip size, shape, continuity, and chip serration), cutting forces, vibrations caused, hardness, surface integrity, and weldability of chips with the tool and/or workpiece. The elemental composition (percentage of alloying elements) and constituents of the aluminum alloy also contribute significantly to determining the machinability rating of aluminum alloys. Nonabrasive constituents support machinability, while abrasive constituents impart negative effects, such as deteriorating tool life and surface integrity. Nonabrasive constituents assist in chip breakage and lead to easier machining.

This chapter will present and discuss the problems and solution approaches associated with the milling of aluminum alloys in light of research conducted by various scholars. The first part of the chapter includes the following aspects: a brief classification of aluminum alloys, types of milling operations and cutting tools, cutting forces, residual stresses, heat generation and temperature distribution, dimensional variation, burr and chip formation, roughness of machined surfaces, tool wear rate, and premature tool failure. The second part of the chapter addresses approaches to overcoming the milling challenges in aluminum alloys. These approaches include tool coatings, parametric optimization, minimum quantity lubrication (MQL), and energy-assisted milling (EAM) (ultrasonic-assisted milling (UAM), plasma-assisted milling (PAM), laser-assisted milling (LAMill), etc.).

2. Brief classification of aluminum alloys

Approximately 8% of the Earth's crust consists of aluminum, making it the third most abundant element [2]. With reference to its crystal structure, aluminum has a face-centered cubic crystal structure with a low melting temperature of 659°C (1,220°F) [3]. Aluminum and its alloys have comparatively higher values of strength-to-weight ratio, corrosion resistance, and conductivities (thermal as well as electrical). Additionally, the aluminum alloys offer nonmagnetic behavior along with good formability and machinability [4, 5]. Aluminum alloys are classified into eight series developed by the Aluminum Association, as shown in **Table 1**.

Nearly 75% of the aluminum produced is still in use, which demonstrates its durability and recyclability [6]. A wide range of applications is reported in the literature where aluminum and its alloys are utilized. These include packaging applications, architectural and structural applications, automobiles and aerospace applications, high-voltage transmission lines, and many others [7–9]. Aluminum alloys make up 70–80% of the total weight of fighter aircraft and 60–70% of the total weight of passenger airbuses [10, 11].

In general, aluminum and its alloys are considered soft alloys and are easy to machine. However, there is a wide range of aluminum alloys, including both soft and hard alloys. Among the different series of aluminum alloys, the 2xxx series (such as 2024, 2618) and the 7xxx series (such as 7,075, 7,475, 7,050) are considered hard aluminum alloys and offer high strength and toughness. Those alloys that contain more than 10% silicon are known as difficult-to-machine alloys. The presence of hard particles of silicon poses challenges, leading to early tool failure and a low surface finish of the machined features. Some of the alloys from this series (especially if they contain lead in their chemical composition) are easy-to-machine, for example, AA 2011.

Series no.	Alloying elements	Common alloys	Alloy category	Properties
1 xxx	Aluminum	1050; 1060; 1100; 1145; 1200; 1230; 1350	Commercially pure	Excellent corrosion resistance, high thermal and electrical conductivity, low mechanical properties, and excellent workability
2 xxx	Copper and magnesium	2011, 2014, 2017, 2024, 2124, 2219	Heat treatable	High strength and toughness, reduced corrosion resistance, and good creep resistance at high temperatures
3 xxx	Manganese	3003, 3004, 3105,	Nonheat treatable	Moderate strength and good workability
4 xxx	Silicon	4032, 4043, 4145, 4643	Nonheat treatable	Low melting point, wear resistance, and usage in automotive applications such as engine parts
5 xxx	Magnesium	5005, 5052, 5083, 5086, 5182	Nonheat treatable	Moderate-to-high-strength, work-hardenable alloy with high corrosion resistance, good weldability, and machinability
6 xxx	Magnesium and Silicon	6061, 6063	Heat treatable	Medium strength with good formability, weldability, machinability, and excellent corrosion resistance
7 xxx	Zinc	7075, 7050, 7049, 710.0; 711.0	Heat treatable	Moderate-high strength, reduced resistance to stress, corrosion resistance, and resistance to cracking and fracture toughness
8 xxx	Others, for example, lithium and tin	8006, 8111, 8079, 8090	Nonheat treatable	Aluminum–lithium alloy 8090, which has exceptionally high strength and stiffness

Table 1.
Aluminum alloys and their properties.

3. Types of milling operations and cutters

Milling is a process in which the workpiece is fed against the rotating tool (the milling cutter). The cutter usually has multipoint cutting edges; however, single-point cutting tools are also used. The milling process is equally suited to machining flat and curved surfaces [12]. Each cutting edge contributes to the removal mechanism and subtracts material from the workpiece in the form of chips during tool–material interaction [9]. There are several types of milling operations practiced in industrial and research settings, as depicted in the upper part of **Figure 1**.

Peripheral milling, also called plain milling, is the type in which the tool's axis of rotation remains parallel to the workpiece orientation. On the other hand, when the tool's axis of rotation is perpendicular to the workpiece, the milling type is called face milling. A similar principle applies to the end milling operation. When two (or more) cutters are mounted on the arbor to remove material from two (or more) surfaces simultaneously, the operation is termed straddle milling. With the help of specialty-formed cutting edges (form tools), customized curved milling is performed, and the corresponding type of operation is called form milling. Slotting is another type of milling operation used to produce grooves and slots. Different splines and keyways are examples of slotting operations. Similarly, slitting is used for parting off the workpiece

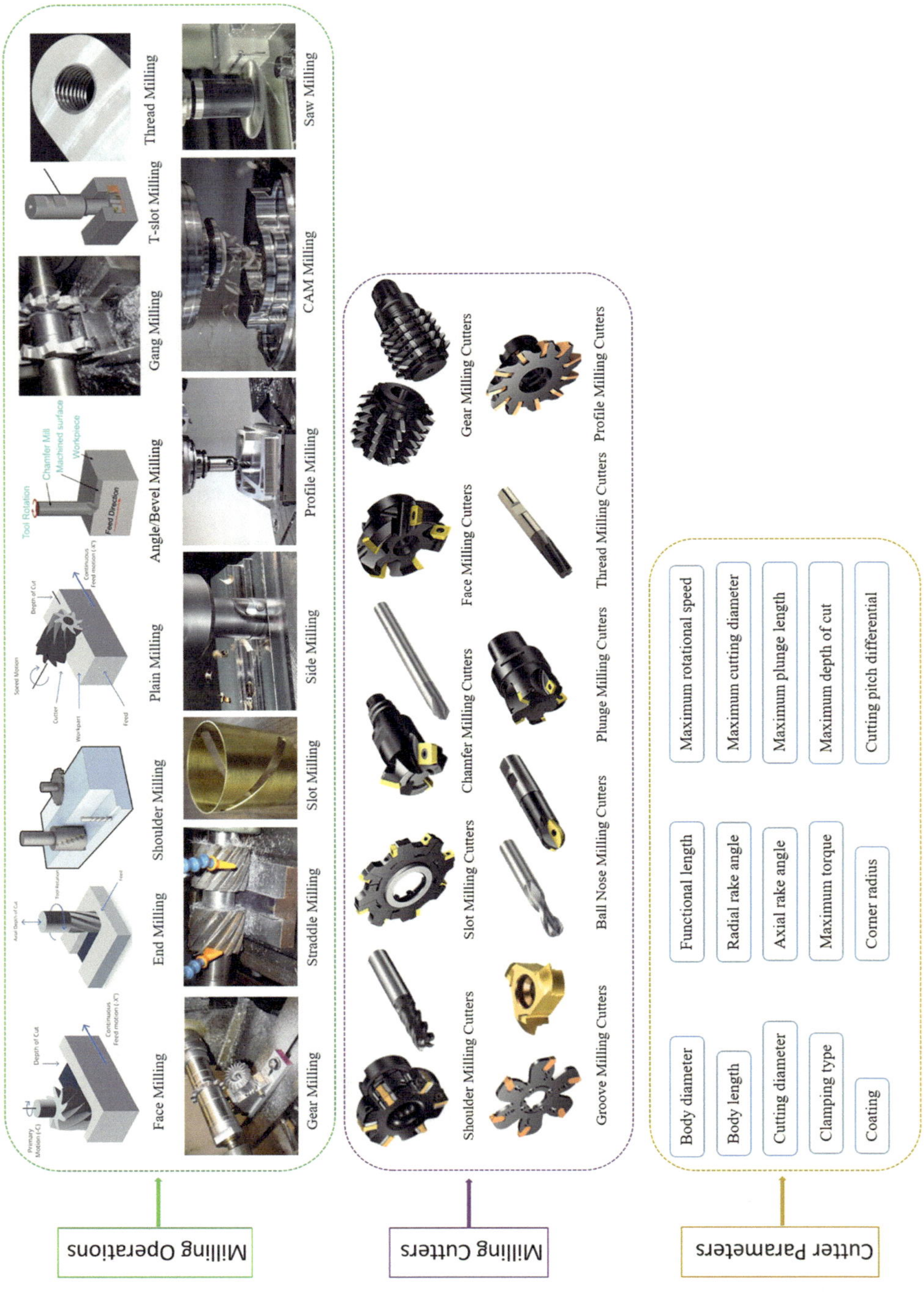

Figure 1.
Different types of milling operations, cutters, and cutter parameters [13].

[2]. Other milling operations include shoulder milling, angle/bevel milling, T-slot milling, gang milling, thread milling, gear milling, CAM milling, and saw milling.

Based on the type of milling operation, the milling cutter is named and selected to perform the desired operation. For example, the cutter used to machine slots is termed as a slot milling cutter; thread milling cutters are used to generate different types of threads; the tool used to develop shoulders in the workpiece block is called a shoulder milling cutter; gear milling cutters are specialty tools used to machine different gear profiles; and the tools used to perform face milling operations are called face milling cutters. All such tools are mounted in the tool holder and/or on the arbor connected to the machine spindle.

There are different ways to classify milling tools and cutters. However, one classification can be based on the attachment of the tool inserts. In the majority of milling operations, the tool inserts are mounted on the cutter body; for example, face, slot, and plunge milling cutters have the tool inserts mounted on the main body of the cutter. It can be a single or multiple tool inserts attached to the cutter body, which is mounted on the tool holder, as shown in the middle part of **Figure 1**. The advantage of this class of milling cutters is that only the tool insert undergoes wear, and the whole cutter doesn't need to be changed upon wear. On the other hand, some milling cutters act as a single-unit tool wherein no insert is attached; instead, the cutter itself performs the milling operation. Gear milling cutters, form cutters, and thread milling cutters are examples of such tools. The wear in such tools results in the replacement of the complete tool, and consequently, the machining cost increases.

In each category of milling cutters, there are several choices to consider for selection in terms of the variety of tool materials, coated and uncoated tools, and the range of manufacturers. However, cutters have numerous specifications and parameters that should be taken into account when making a tool selection. Some of the important parameters of the cutters include cutting diameter, functional length, corner radius, rake angle (radial and axial), maximum rotational speed, and maximum depth of cut. The frequently used cutter parameters are listed in the lower part of **Figure 1**. With reference to the selection of a suitable cutting tool for the desired milling operation, several factors are involved, such as the type of material, alloying elements, milling profile, milling depth, surface finish, machining time, and cutting parameters. Therefore, determining the most suitable tool material and its specific type is a complex task.

4. Challenges in milling of aluminum alloys

Milling of aluminum and its alloys also poses challenges depending on the type of alloy and its constituents. For example, alloys in the 2xxx and 7xxx series are still considered challenging. In this section, different characteristics of milling aluminum alloys and the associated challenges are presented, such as cutting forces, surface roughness, residual stresses, temperature distribution, dimensional inaccuracies, tool wear, and tool failure.

4.1 Cutting forces

Among the various outcomes of material removal processes, the cutting force is one of the most important concerns. Tool deflection error is also affected by cutting forces, which means that the factors influencing cutting force have notable effects on tool deflection error [14]. Generally, cutting forces are measured in three directions, that is, tangential (F_X), radial (F_Y), and axial (F_Z) directions, by placing a dynamometer beneath the workpiece, as shown in **Figure 2(a)** [15]. The actual experimental setups used to measure the cutting forces while milling aluminum alloys are shown in **Figure 2(b** and **c)** [16].

Cutting forces are low in aluminum alloys compared to steel [3]. These forces depend on a large number of factors, including the workpiece material composition, tool material and its geometry, and the milling parameters [17]. The cross-sectional area of the chip and the cutting forces create the cutting force coefficients (shear and edge). These coefficients are functions of cutting depth. Under the effect of cutting depth, the shearing coefficient shows a large variation, while the edge coefficient undergoes a relatively small variation, as shown in **Figure 3** [18]. Campatelli et al. [19] reported that the cutting speed and feed rate, along with the chip thickness, influence the shearing coefficient. However, the effect of cutting speed is comparatively insignificant compared to instantaneous chip thickness and feed rate.

The cutting force is largely affected by the spindle speed (~92%) and feed rate (~6%), as reported in [20], but the F_Z component contributes less to the cutting force compared to F_X and F_Y [17]. The axial depth of cut and the feed per tooth significantly affect the tangential and radial cutting forces, as reported in [21], wherein the ball-end milling of AA2014-T6 was performed. According to Subramanian et al. [22], the minimum cutting force can be achieved by setting the

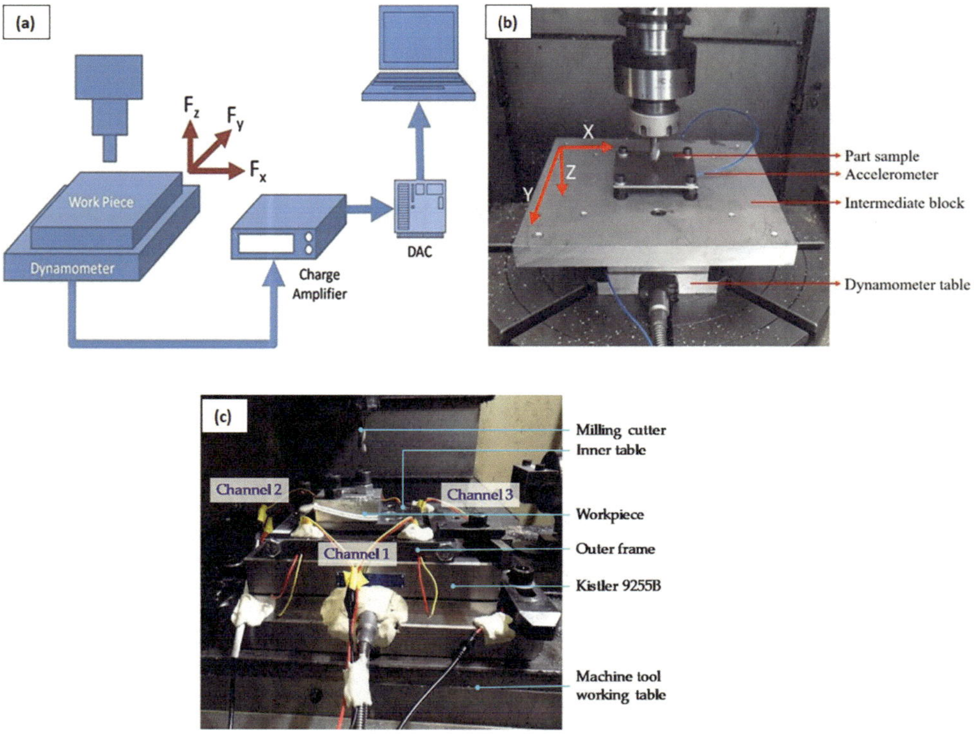

Figure 2.
Cutting force measurement: (a) schematic diagram [15], (b) experimental setup of 2024-T3, and (c) experimental setup for aluminum alloy [16].

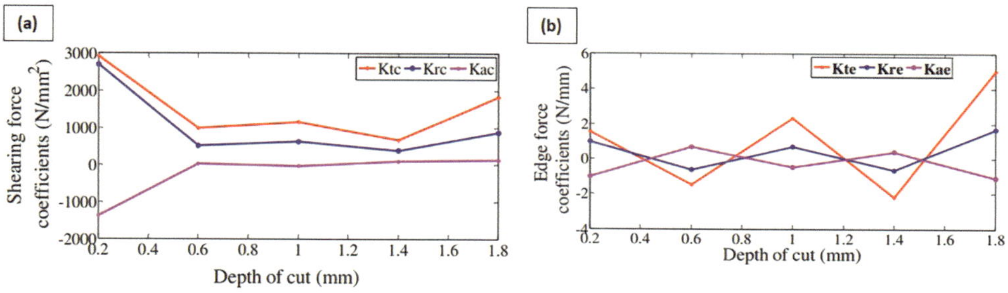

Figure 3.
(a) Shearing coefficients vs. depth of cut and (b) edge coefficients vs. depth of cut [18].

cutting speed at 100–160 m/min, the depth of cut at 0.5–1.5 mm, and the feed rate at 0.06–0.09 mm/tooth.

In milling, the thrust force is perpendicular to the cutting force. According to Niu et al. [23], in the case of the tungsten carbide tool, the cutting and thrust forces are equal to each other, and with the use of a natural diamond tool, the cutting force is double the thrust force. Compared with both of these materials, the tungsten carbide tool undergoes higher forces. Fine-grain diamond coatings offer an anti-adhering characteristic with a low friction coefficient (0.1). Hence, the performance of tungsten carbide tools can be enhanced by using fine-grain diamond [15, 24]. Generally, a smaller positive rake angle results in higher cutting forces due to the reduction in shear angle. For a polycrystalline diamond (PCD) cutting tool, the resultant forces are lower [25]. Thus, increased cutting speed, reduced feed rate, and a low amount of depth of cut contribute to lowering the cutting forces in aluminum alloys.

4.2 Residual stresses

Residual stresses are produced due to thermal, mechanical, and structural loads (including phase transitions under the impact of stresses) [26]. Typically, thermal and mechanical loads generate tensile and compressive residual stresses, respectively [27]. However, thermal load is the dominant factor influencing residual stresses. During the milling of aluminum alloys, residual stresses are induced by cutting forces, which can impact the shape, dimensional accuracy, and properties of the milled part [28]. These stresses directly affect the component's lifespan in terms of creep and cracking resistance. Residual stresses are primarily influenced by cutting parameters and tool path strategies (single-pass and multipass), as observed in the milling of aluminum alloy 7050-T7451 [29] and 7050-T6 [30]. The residual stress depth profiles of two different grades of Al 7075 have been studied in [31], as illustrated in **Figure 4**. An increase in cutting speed or feed per tooth results in higher compressive residual stresses [32]. Residual stresses in aluminum alloys are generally compressive but tend to become tensile as the cutting speed increases [33, 34]. However, during face milling of AA7050-T7451, tensile surface residual stresses were observed at low cutting speeds (200 m/min) due to higher thermal effects, while compressive residual stresses were formed at higher cutting speeds (800 m/min) due to increased machining forces [35]. The flank wear of the tool also significantly affects the propagation of residual stresses. Tensile stresses decrease

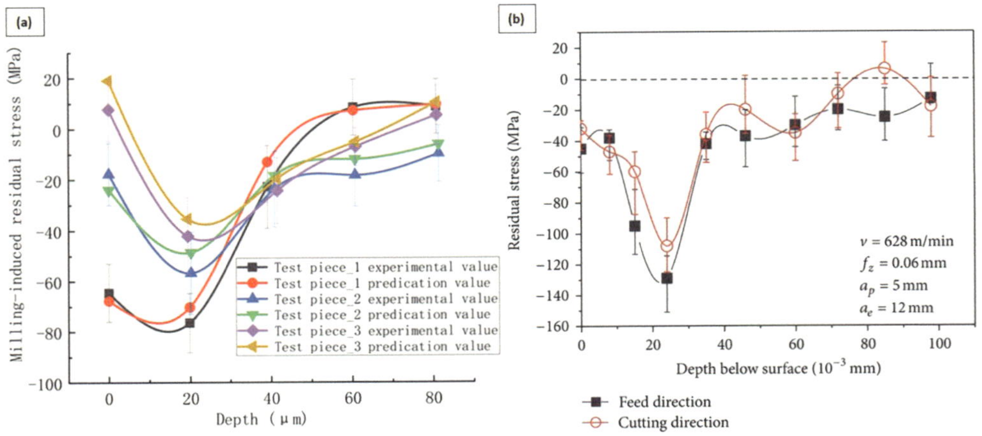

Figure 4.
Residual stress depth profile in milling: (a) Al 7075 [31] and (b) Al 7075-T7451.

when flank wear is high, particularly at low cutting speeds (100 m/min), and stresses increase with greater flank wear at high cutting speeds [36]. Tang et al. [37] reported that during the milling of AA7050-T7451, tensile or compressive stresses are minimal at the initial stages of flank wear, but compressive stresses transition to tensile stresses as flank wear progresses.

4.3 Heat generation and temperature distribution

Most of the heat (99%) transfers to the tool, substrate, and machining chips [3]. Among these three constituents, a high proportion of heat is transferred to the workpiece, which leads to thermal expansion and residual stresses. Thermal expansion, in turn, produces dimensional and geometrical variations. Thermal expansion and dimensional variation due to heat generation are the major issues of low-rigidity aluminum alloys [38]. The temperature in milling is measured with the help of thermocouples. Heat generation in aluminum alloys is mostly affected by the shape of the cutting tool and the cutting parameters. The maximum temperature occurs at the tool–workpiece interaction point. The tool edge radius (TER) has a significant impact on temperature distribution [39]. An increase in the TER reduces the overall cutting temperature, as shown in **Figure 5**.

Other cutting parameters also contribute toward the amount of heat generation [40]. Feed and cutting speed have an inverse relationship with heat generation in the milling of AA2024-T3 aluminum alloy [41], but feed has more influence than cutting speed. Heat generation is also linked to the machining profile, that is, smoother curves generate less heat compared to sharp trajectories [27]. In a study on the milling of AA-6063 [42], it was stated that the depth of cut contributes 86.2% to the workpiece temperature rise. The use of a chamfered-edge cutting tool increases stability but also raises the risk of soft spot formation due to greater heat generation compared to a sharp-edge cutting tool. The maximum temperature generated in Al–Li 2196 for a sharp-edge cutting tool is less than 100°C, whereas the maximum temperature generated for a chamfered-edge cutting tool can reach up to 300°C [43]. With an increase in the feed rate per tooth, the maximum temperature in the chamfered mill can be reduced to 150°C, as shown in **Figure 6**. Moreover, the

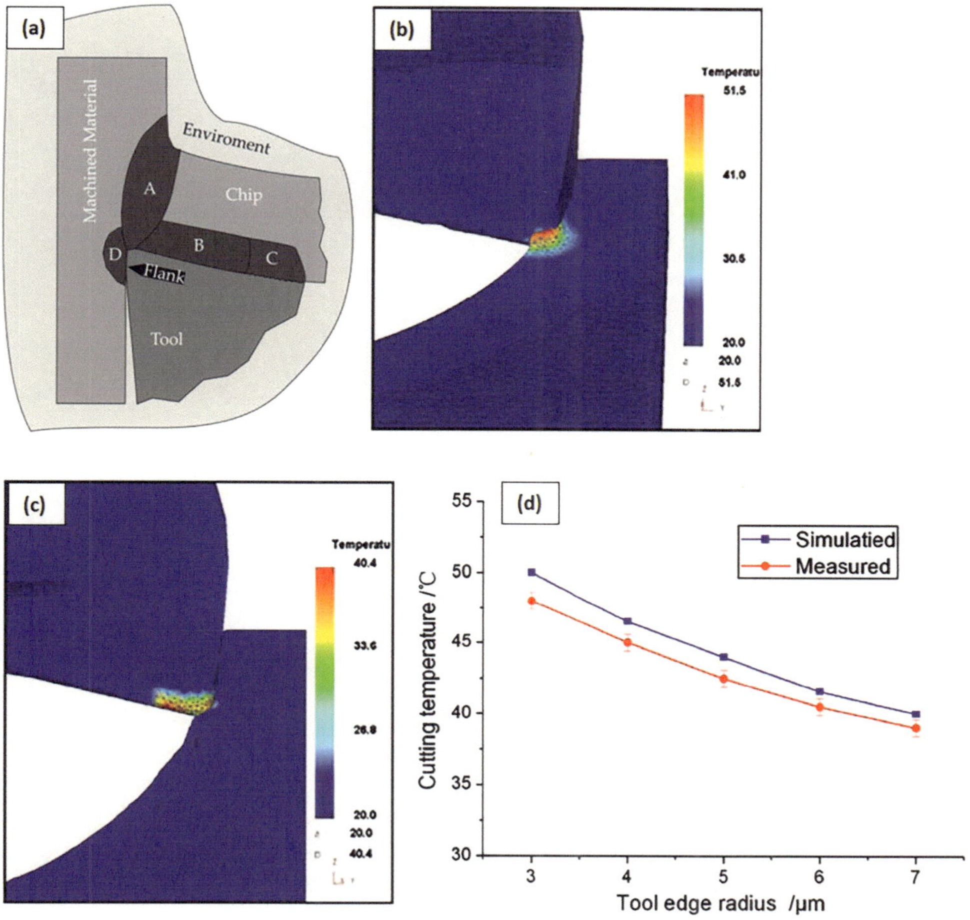

Figure 5.
Heat generation in milling: (a) schematic of heat generation zones, (b) heat generation with TER 3.2 μm, (c) heat generation with TER 7.0 μm, (d) effect of TER on cutting temperature [39].

amount of accumulated heat also depends on the milling time, especially at the beginning of the machining cycle.

4.4 Dimensional variation

Owing to their higher thermal conductivity, the heat dissipation in aluminum alloys is very quick, and the overall temperature of the workpiece rises in a very short time. With a change in temperature, the size of the workpiece expands (during cutting) and contracts (during cooling). This brings difficulties in controlling the dimensional accuracy of aluminum alloys [10]. In order to maintain the desirable temperature, cutting fluids with high coolant ability can be used [44]. In an aluminum workpiece with a hardness of 21 HRB, dimensional variation can be controlled by using higher feed rates and adequate coolant [45]. In another study [46], it is reported that the tool holder also contributes to dimensional variations in aluminum parts during the milling operation. The dimensional variations and geometrical inaccuracies in aluminum alloys are more evident and challenging when

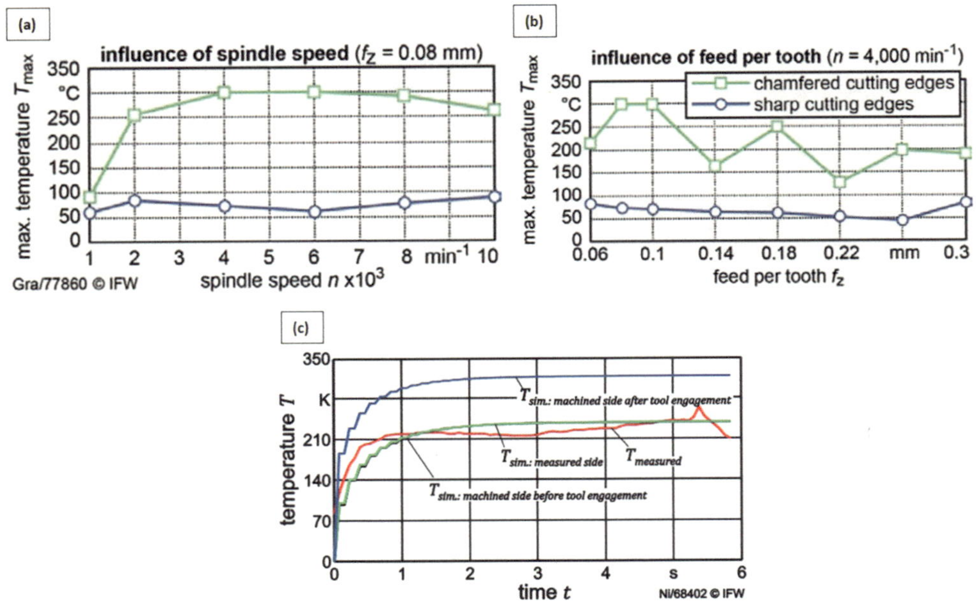

Figure 6.
Effect of parameters on cutting temperature: (a) spindle speed, (b) feed per tooth, and (c) milling time [43].

producing thin-walled structures [47]. The dimensional variation depends on the elemental composition of the alloy, the geometry of the thin-walled features, as well as the type of milling operation, as shown in **Figure** 7. Different structures (semi-open structures and closed structures) undergo different levels of deformation. A comparison of high-speed milling (HSM) and conventional milling has been carried out in Zawada-Michałowska et al. (2022) [48] to examine the dimensional variations in different structures.

Deformations in aluminum also depend on its alloys. For example, the AW-7075 T651 alloy undergoes relatively more deformations compared with AW-6082 T651 and AC-43000. The milling of AA6063-T6 under dry conditions shows that dimensional variation is largely affected by spindle speed (53.05%), followed by depth of cut (26.13%) and feed rate (3.51%), whereas material removal rate (MRR) is largely affected by the depth of cut (63.10%) and the feed rate (33.30%). Optimum parameters for dimensional variation of AA6063-T6 under dry conditions are reported as a spindle speed of 2000 rpm, a depth of cut of 0.048 mm, and a feed rate of 1.8 mm/min [49]. Thus, low spindle speed and depth of cut, and high feed rate with adequate coolant are recommended for better control over temperature rise and dimensional variations.

4.5 Burr and chip formation

Burrs are formed as the tool starts to wear and become larger with increased tool wear [50]. Burrs are governed by tool wear, undeformed chip thickness, cutting speed, friction angle, tool geometry, and feed per tooth. An optimized parametric combination is usually required to minimize burr formation at different regions of the milled slots [51]. Chip control is a major problem in machining aluminum alloys.

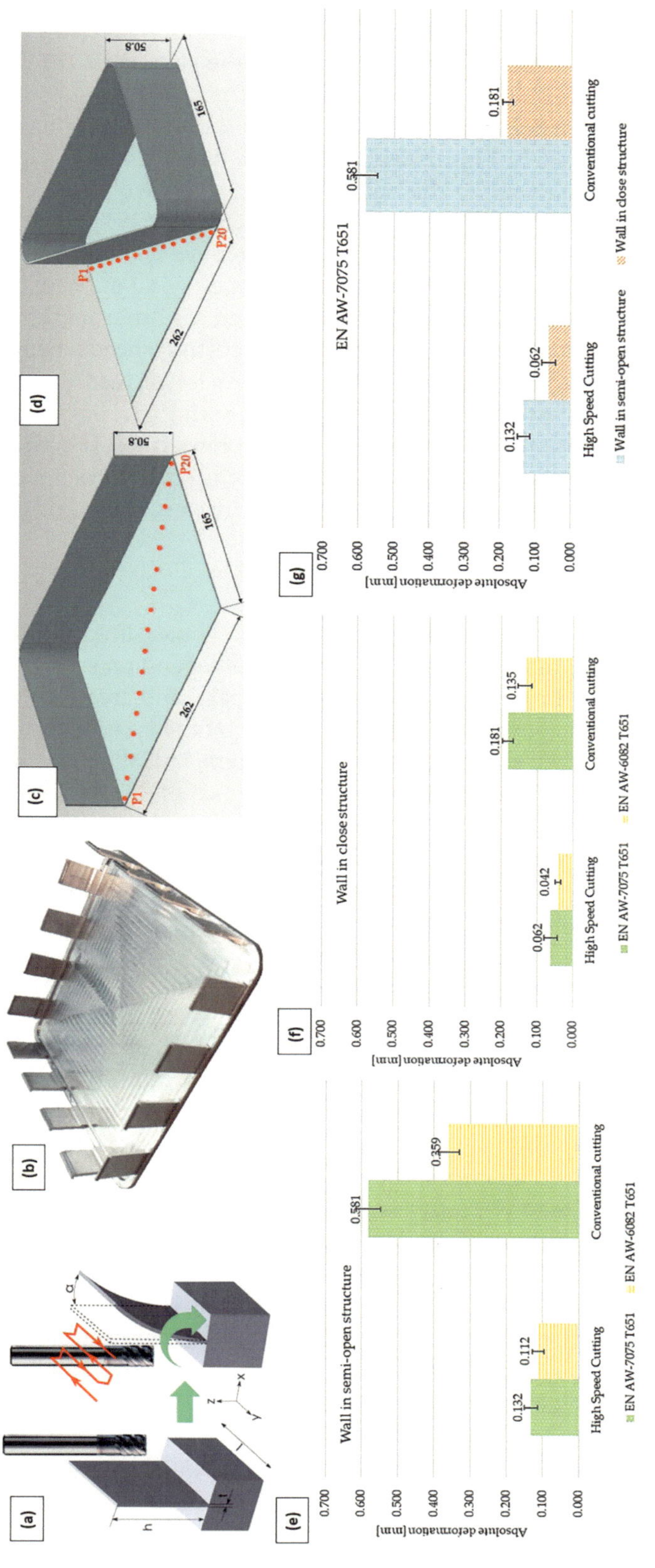

Figure 7.
Dimensional variation in thin-walled features of aluminum alloys: (a) deformation schematic, (b) element of a snap connection in aircraft skin [47], (c) semi-open thin-wall structure, (d) close thin-wall structure, (e–g) deviation in comparison between conventional and HSM [48].

Continuous and thick chips are formed in some aluminum alloys [3]. These continuous chips can form built-up-edge (BUE) on the tool surface, twist around the tool, and enter the micro-channel, interfering with the cutting forces [50]. The formation of burrs at different regions and BUE in the case of different aluminum alloys is shown in **Figure 8**. For instance, the BUE in the case of milling AA2024 [52] and A97050 [53] is shown in **Figure 8(b** and **c)**. The amount and thickness of BUE directly affect the morphology of the milled surface. In controlling the BUE in Al6061, the feed rate and cutting speed play a significant role [54]. Chip formation in aluminum alloys is also affected by tool type and its geometry, feed rate, and material composition. The addition of lead, bismuth, or tin as alloying elements facilitates the breaking of long chips into short chips [55]. There are other methods as well to break down continuous chips, for example, at high cutting speeds using tools with polished teeth and large rake angles [56]. At relatively lower feed rates, the undeformed chip thickness depends on the feed per tooth [57]. Higher feed per tooth and chip thickness reduce the friction angle at the tool interface, thereby resulting in smaller burr thickness [58, 59]. Larger undeformed chip thickness parameters also help in reducing milling energy consumption by decreasing the cutting time [60].

4.6 Surface roughness

The quality of machined parts in industries is usually measured through their surface roughness. Several mechanical properties also depend on surface roughness, such as corrosion resistance, fatigue behavior of the part, and creep life [61]. Surface roughness in aluminum alloys is typically influenced by the type of cutting tool and its geometry, tool path strategy, cutting parameters, and alloy hardness [62]. Cutter diameter and edge radius have a significant impact on the surface roughness of aluminum alloys. For instance, a 0.9-mm-diameter tool produces a smaller roughness value as compared to a tool with a 0.6-mm-diameter, as shown in **Figure 9(a)** [63]. In end milling, a tool with a greater number of flutes can lead to lower surface roughness [64]. Small cutting-edge radius inserts produce less friction at the tool–workpiece interface, resulting in lower cutting forces. Consequently, the surface roughness is reduced compared to a tool with a larger cutting-edge radius, as shown in **Figure 9**.

In most cases, uncoated tools operating at low cutting speeds and coated tools at high cutting speeds provide better surface finish, as shown in **Figure 10(a–d)** [65]. The feed rate and cutting speed interactively have a strong influence on surface roughness during the milling of aluminum alloys and composites. However, the feed rate exerts a comparatively greater influence than cutting speed. High spindle speed, combined with a low feed rate, reduces the surface finish during the milling of aluminum composites such as CFRP-Al laminated composites [66] and SiC-Al metal matrix composites [67]. The effect of increasing the feed rate on the surface of AA-6061 raster-milled material is shown in **Figure 10(e** and **f)** [68].

The tool path strategy plays an important role in determining the milling time and surface roughness. When comparing different tool path strategies, the constant overlap spiral path was found to be the quickest and had a lower roughness value among parallel spiral and parallel zig-zag machining strategies [69]. It has been reported that the tool path strategy contributes more than 70% to achieving low surface roughness and machining time compared to other milling parameters [70]. The raster tool path strategy is reported as the optimum strategy

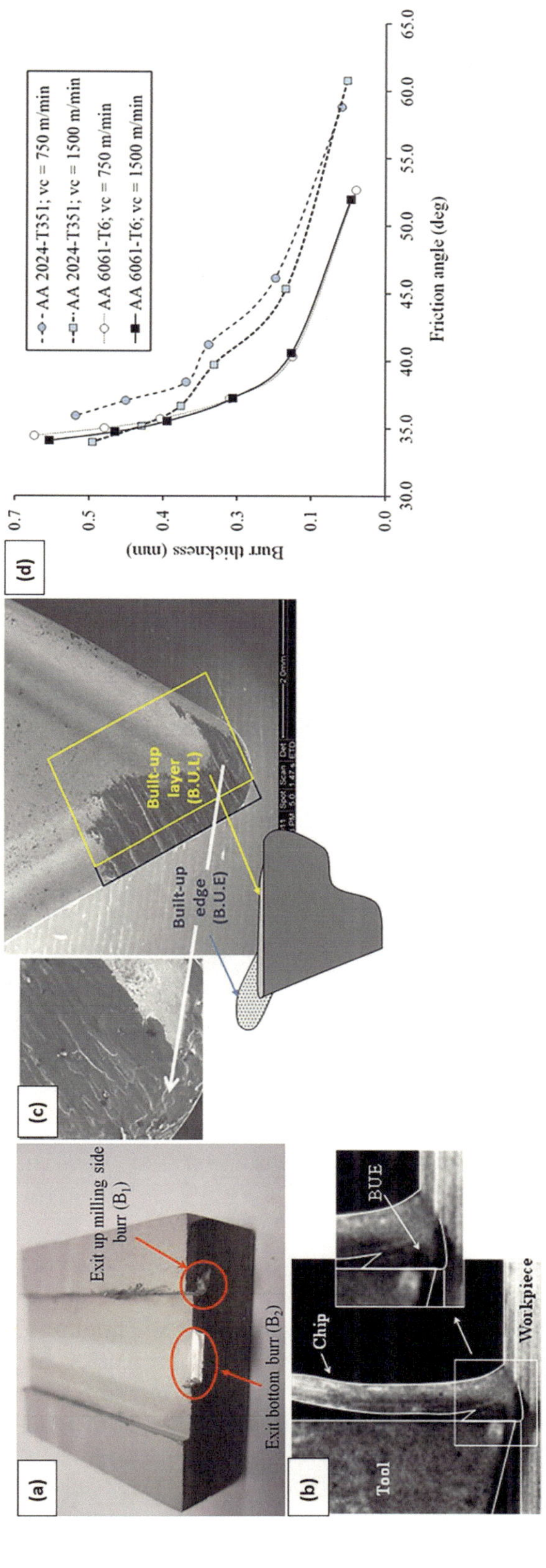

Figure 8.
Burr and BUE formation: (a) slot milling exit burrs [59], (b and c) BUE aluminum alloy AA2024 [52] and A97050 [53], and (d) burr thickness in AA2024 and 6061 [59].

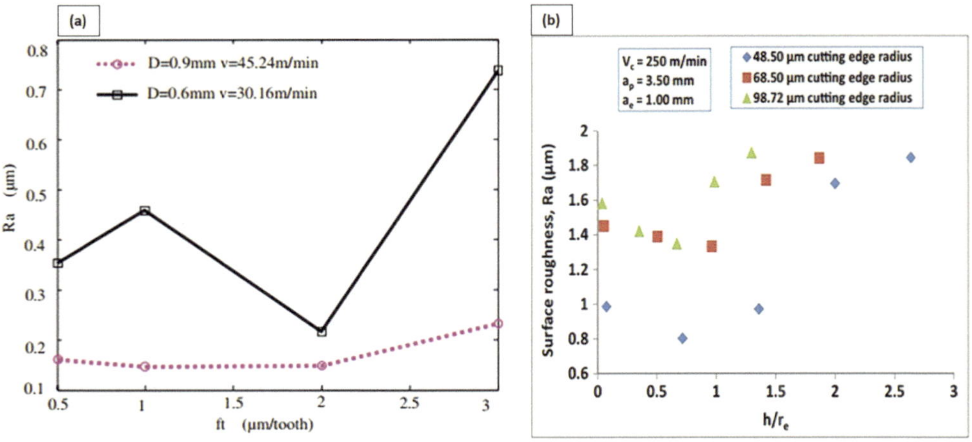

Figure 9.
Roughness as a function of (a) tool diameter [63] and (b) edge radius during milling of AW6082-T6 [57].

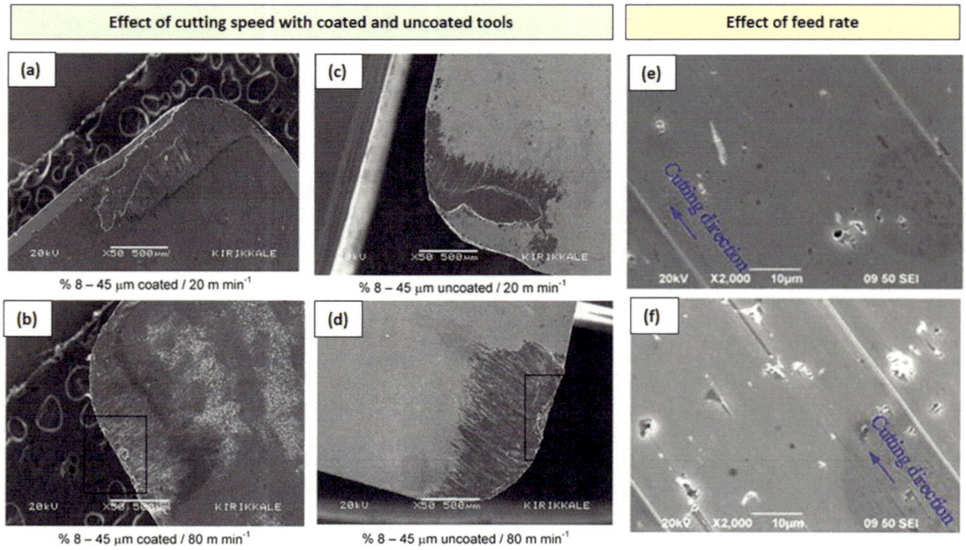

Figure 10.
Effect of milling parameters on roughness: (a–d) worn cutting edges of coated and uncoated tools at 20 and 80 m/min cutting speeds [65], (e) feed rate 50 mm/min, and (f) 300 mm/min [68].

for milling AA-6005, achieving a roughness of 0.43 μm. The cooling mode also contributes toward surface roughness, and cryogenic cooling has been reported to be the most effective mode of cooling compared to flood cooling [71]. The selection of the most appropriate milling conditions depends on the type of milling alloys. However, in general, surface roughness can be minimized for aluminum alloys by selecting suitable cutting tools and associated milling conditions, such as high cutting speeds, diamond tools or diamond-coated cutting tools with small cutting-edge radii, the use of coolant, large rake angles, and low feed rates [65, 72, 73].

4.7 Tool wear

Tool wear in milling operations is unavoidable, resulting from continuous contact between the tool tip and the workpiece material. Generally, flank wear is found in the milling of aluminum alloys, but notch wear is more prominent in hard aluminum alloys [74]. Aluminum alloys containing hard particles (such as Al_2O_3, SiC, and Si >12%) are considered difficult-to-machine alloys. These alloying elements or precipitates increase the abrasive action on cutters [56, 74, 75]. For milling hard aluminum alloys, the preferred choice of tool is PCD tools, followed by carbide tools and coated tungsten carbide (WC) tools. Tools made of high-speed steel (HSS) are usually not recommended for these alloys [25, 75]. Xu et al. [76] preferred a Ti (C7N3)-based cermet tool over a WC tool in micro-end milling of AA-2024. A comparison was made to observe burr formation. It was found that under identical cutting conditions, burr formation was observed after 13 minutes when using a Ti (C_7N_3) tool, whereas with a WC tool, burrs developed within 6 minutes.

The wear mechanism for tungsten carbide tools is edge rounding, while for brittle PCD tools, it is the microfracture of the cutting edge. Smearing phenomena also occur with WC tools due to an increase in the cutting-edge radius, which can be avoided by using PCD tools [25]. Coated WC tools used in ball-end milling of AA6061-T6 typically experience significant notch and flank wear due to abrasive contact with the workpiece. Flaking and cracking of the coating at the cutting edge [77], as well as adherence wear [78], are other types of wear modes observed during the cutting process, as shown in **Figure 11**.

Adhesion wear, microchipping, and diffusion wear are also observed while milling aluminum alloys. Chip adhesion with the cutting tool is another phenomenon in the dry milling of aluminum alloys. It imposes challenges in achieving a high surface finish on the machined surfaces. In a research study [79], this challenge has been addressed by developing dimple-shaped textures on the cutting tool surfaces, as shown in **Figure 12**. The ratio between the concave and convex areas of the texture determines the adhesion on the cutting surface. In comparison with the conventional tool, using a dimple-textured tool, the amount of chip adhesion at the cutting surface of the tool can be reduced to 10%.

4.8 Tool failure

The ultimate impact of tool wear is the deterioration of the surface finish of the milled material and the premature failure of cutting tools. The premature failure

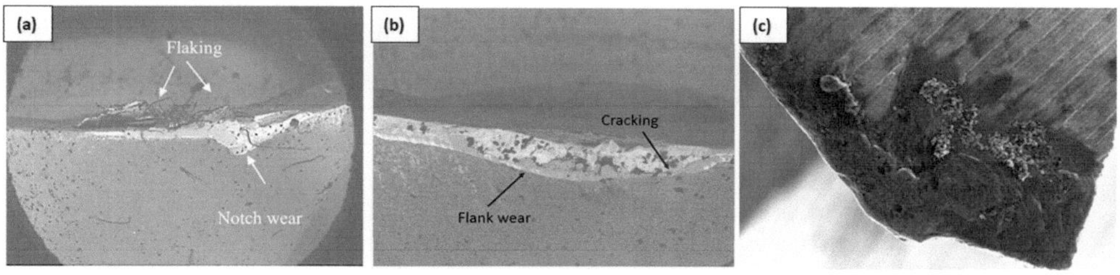

Figure 11.
Tool wear modes: (a) flank and notch wear in end milling of aluminum 6061, (b) cracking and flank wear on the cutting edge [77], and (c) adherence wear on the cutting edge [78].

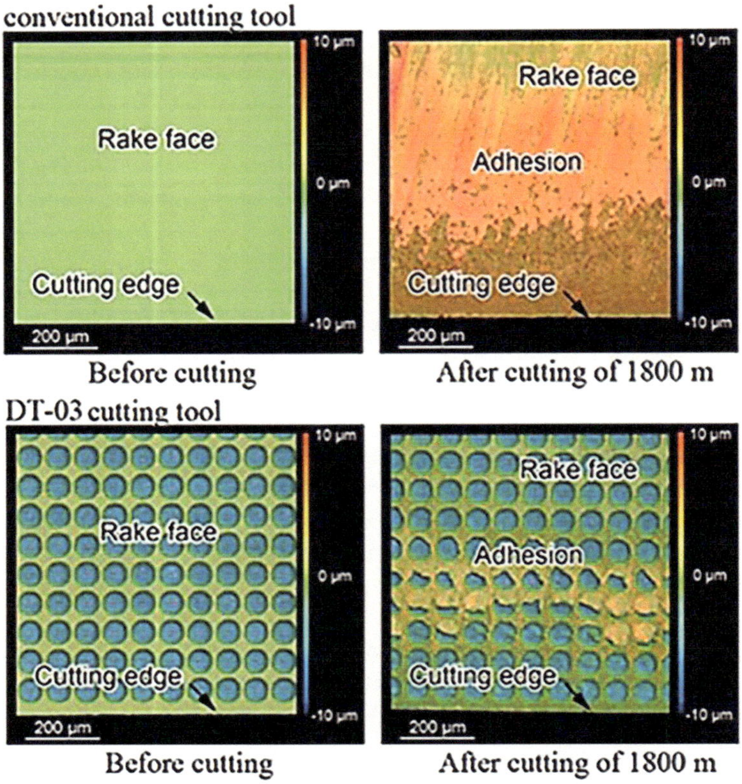

Figure 12.
Rake face of conventional and dimple-textured (DT-03) cutting tools before and after cutting [79].

mechanism is common in HSM of aluminum alloys. In addition to tool wear, the reduction in fatigue strength, stress concentration, notching effect, dynamic imbalance, and vibrations due to changes in cutting forces are the main reasons for premature tool failures [10]. Calatoru et al. [78] found catastrophic wear and failure of tungsten carbide with cobalt binding (WC–Co) tools during HSM of aluminum alloys. Due to strong heat generation in HSM, aluminum reacted with cobalt, forming cobalt aluminide, which left the carbide without sufficient bonding, resulting in spontaneous tool failure. To effectively mitigate early failure in high-speed cutters, materials with high impact and wear resistance – such as PCD and WC – should be employed. Thermal fatigue is another characteristic that plays an important role in the selection of appropriate tool materials. The use of ultrafine-grain carbide or fine-grain matrix materials in cutting tools can significantly reduce the likelihood of early failure. Moreover, some high wear-resistant coatings can withstand longer durations, reducing the chance of early tool failure and increasing tool life.

5. Overcoming milling challenges in aluminum

There are several approaches adopted to address the above-mentioned challenges encountered during the milling of aluminum and its alloys. Such approaches include the use of appropriate tool materials, the use of coated tools, optimization of milling parameters, the combination of coated tools with optimized parameters, MQL, and external/additional EAM.

5.1 Tool coatings

The coating of the cutting tool plays an important role in reducing tool wear and the chances of early tool failures. Excessive heat generation is inevitable in milling, especially HSM. Under the effect of high heat generation, aluminum may react with the coating material, as well as tend to weld with the cutting edge, generating an adhesion phenomenon. Therefore, the selection of the appropriate coating material is crucial, especially during the dry milling of aluminum alloys. Hence, the coating must be applied to tools containing elements with low affinity for aluminum to avoid these reactive and adhesion phenomena. For example, diamond-like coatings (DLC) and diamond-coated tools exhibit exceptionally low friction coefficients and minimal affinity toward aluminum alloys [80], outperforming all other hard coatings such as TiC, TiN, TiB2,TiB_2, Al2O3,Al_2O_3, and AlON [74]. To optimize dry milling of aluminum alloys, it is advisable to use coatings such as WC/C, CrC/C, or TiN + MoS_2 with soft-phase morphology, as well as superhard diamond coatings, for enhanced wear resistance and reduced friction [81]. Soft composite coatings are used for inferior cutting conditions compared to diamond coatings. Soft composite coating Graphite-PTFE (Gr–x), as compared to MoS2 on HSS tools, is inexpensive and exhibits nonwetting characteristics toward aluminum alloys [82].

5.2 Optimal parameters

Appropriate tool coating has the potential to solve several cases of tool wear. However, in many cases, it is necessary to perform parametric optimization in order to achieve the desired milling results while dealing with aluminum alloys. Sometimes, the combination of the tool coating and the optimal parameters is needed together. Notably, TiAlN coating is normally inappropriate in the milling of aluminum alloys because it produces a chemical reaction with aluminum [10]. However, the selection of optimal machining parameters can prevent this reactive phenomenon with aluminum. Cutting speed, feed rate, and appropriate depth of cut are the most important parameters among others because these parameters contribute significantly to heat generation. For instance, during dry milling, the life of a tungsten carbide tool coated with TiAlN can be increased from 11 to 97 minutes. Cutting speed plays a pivotal role in determining the lifespan of tungsten carbide cutting tools, with feed rate and depth of cut being secondary factors. Optimal tool life (up to 97 minutes) can be achieved using a cutting speed of 115 m/min, a feed rate of 0.15 mm/tooth, and a depth of cut of 0.5 mm [77].

To achieve improved milling performance with coated tools, selecting the optimal combination of machining parameters is essential. Tsao et al. [83] employed a WC tool with TiAlN + DLC coating to perform milling of AA6061P-T651. Under the recommended optimal combination of milling parameters, a significant decrease in flank wear (from 0.177 mm to 0.067 mm) and surface roughness (from 0.44 μm to 0.24 μm) was observed. The recommended optimal milling conditions include a 60° relief angle, 10 mm cutter diameter, 2.5 mm depth of cut, 3 mm cutting width, 0.02 mm/rev feed rate, and 5500 rpm spindle speed. Tool wear during micromilling of

AA-7075 using CB3A 20080–040 ball end mills increases with spindle speed and depth of cut but decreases with an increased feed rate. The optimal parameters for minimal wear are a 10,000 rpm spindle speed, 0.5 μm/tooth feed rate, and 50 μm depth of cut [84].

5.3 Minimal quantity lubrication (MQL)

In milling, lubrication helps dissipate heat, reduce tool–workpiece friction, and remove chips from the cutting zone [85, 86]. The cost associated with cutting fluids often exceeds labor and overhead expenses by a significant margin [87]. The use of lubrication represents 20% of the machining costs [88]. From the perspectives of economic efficiency, environmental sustainability, and occupational health, MQL is recommended over traditional cooling methods due to its minimal effect on multiple performance metrics. The MQL technique involves the atomization of cutting fluid mixed with air. By reducing friction at the cutting interface, oil droplets in MQL help control temperature, while the air stream contributes to efficient heat dispersion during machining [89]. To balance lubrication efficiency and fluid consumption, MQL systems regulate flow rates between 50 and 500 ml/h. Commonly used cutting fluids for different materials, including aluminum and its alloys, are shown in **Table 2** [90].

The performance of the MQL technique further depends on various controllable factors such as nozzle positioning, fluid type, and flow rate. The MQL technique can reduce the quantity of cutting oil by up to 95% and results in less flank wear compared to conventional lubrication during the milling of AA5083-H112. Nozzle position affects the MQL flow, and the optimum nozzle position is 135° relative to the feed direction [91]. MQL significantly influences the MRR and surface finish in aluminum alloy machining, such as AA-7075 [92, 93]. The performance of uncoated carbide inserts can be enhanced by using MQL [94]. An increase in flow rate in MQL decreases the cutting forces during the milling of aluminum alloys 7075, 6061, and 2024, and has a more prominent effect on surface roughness at low cutting speeds [95]. A better surface finish (R_a = 0.602 μm) for the milling of AA-6063 can be achieved by using 40% lubricant [96].

Mixing nanoparticles (SiO_2, TiO_2, and MoS_2) with MQL further facilitates milling operations. Efficient fluid access to the cutting zone enables significant reductions in both cutting forces and surface roughness, improving overall machining performance [92, 97, 98]. However, higher concentrations (more than 0.5% wt) increase the impregnation of nanoparticles in surface pores, which reduces surface

Process	Tool	Cast iron	Aluminum alloys	Magnesium alloys
Turning	HSS	E1, C1	E1, C1, Sp	O1, Sp
	Carbide	E1, C1	E1, C1	O1, Sp
Milling	HSS	E1, C1	O1, Sp	O1, Sp
Drilling	Carbide	E1, C1	O1, Sp	O1, Sp
Tapping	HSS	E1, C1	O1, Sp	O1, Sp

Code: O1 – mineral oil or synthetic oil, C1 – chemical fluid.
E1 – mineral oil emulsion, Sp – specially formulated fluids.

Table 2.
Commonly used fluids for different materials [90].

finish [98]. According to Karim et al. [99], the mixing of multiwall carbon nanotubes into vegetable oil-based lubricants significantly enhances machining performance for AA-6061, yielding reductions in cutting force (57%), temperature (23%), and surface roughness (32%) [99]. Under aggressive machining conditions involving increased feed rates and depths of cut, the presence of nanoparticles helps prevent tool wear, edge chipping, and adhesion-related failures [100].

5.4 Energy-assisted milling

Energy-assisted milling (EAM) or external energy-assisted milling (EEAM) is a comparatively modern technique used in the milling process. In addition to the conventionally available machining energy in the milling process, an external source of energy is supplied near the neighboring areas of the tool–workpiece. The workpiece area near the tool–work interaction zone is softened through the application of an external source of energy, such as a laser beam. The softened part of the substrate facilitates reducing the cutting forces and consequently lowers the potential of tool failure or excessive tool wear. It also reduces tool clogging, especially when dealing with soft aluminum alloys, and improves the MRR. Poor surface finish, tool wear, residual stresses, dimensional inaccuracies, chip formation, and adhesion of chips with the tool are common issues with the conventional milling of aluminum alloys. This technique is widely used to mill numerous types of engineering materials, especially difficult-to-machine materials. However, the use of EAM in the case of aluminum alloys is not as widespread as it is for other hard materials such as Inconel alloys and titanium alloys. Based on the type of external assisted energy source, the processes are classified as ultrasonic vibrations, laser heating, plasma heating.

5.4.1 Ultrasonic-assisted milling (UAM)

Ultrasonic-assisted milling involves the application of micro-scale high-frequency vibrations to either the cutting tool or the workpiece. It effectively reduces cutting forces and the deformation zone while simultaneously enhancing tool life, MRR, and surface integrity [101, 102]. A comparison of conventional and UAM in terms of important performance metrics is shown in **Table 3**. Many researchers have referred to UAM as ultrasonic vibration-assisted milling (VAM) [103–105]. An ultrasonic power (UP) setup is required to produce vibrations. The setup primarily

Performance measure	Conventional milling	UAM
Cutting force	High	20–40% lower
Surface roughness (R_a)	Moderate/high	Up to 50% better finish
Tool wear	Faster	2–3 × tool life improvement
Burr formation	Significant	Significant reduction
Heat generation	Higher	Reduced due to intermittent cutting
MRR	Limited by heat/wear	Increased in many studies

Table 3.
Conventional milling vs. UAM in terms of performance measures.

comprises an ultrasonic generator, a piezoelectric transducer, a booster, and a horn, which collectively serve to generate and amplify high-frequency vibrations. Two types of vibrations are usually produced, namely torsional vibrations and longitudinal vibrations. Either the tool or the workpiece is vibrated through this setup to facilitate the milling operation and achieve better results. Based on the components and the process of UAM, a flowchart has been developed for better understanding, as shown in **Figure 13(a)**. The schematic diagram of UAM presented in [106] is shown in **Figure 13(b)**.

Several researchers have concluded that the assistance of ultrasonic vibration in the milling operation of aluminum alloys reduces cutting forces and surface roughness [107, 108]. UAM's surface roughness for the nonhardenable aluminum alloy AW-5083 is lower than that of conventional milling, especially at lower spindle speeds. Low surface roughness, as well as a better material proportion profile at a spindle speed of 6000 rpm, can be achieved regardless of the feed rate, as shown in **Figure 14(a)** [109]. In UAM, the maximum cutting force is influenced by spindle speed, feed per tooth, and vibration amplitude. As illustrated in **Figure 14(b)**, the cutting force decreases with increasing UP percentage and vibration amplitude [103].

Higher feed rates and axial depths result in larger uncut chip cross-sections (width and thickness), thereby increasing cutting forces and mechanical load on the tool [110]. An increase in ultrasonic vibration amplitude in the feed direction results in less cutting force, smaller chip formation, shorter cutting time, reduced heat accumulation, and longer tool life [103, 104]. However, excessively high ultrasonic vibration amplitudes may induce mechanical shock and tool instability, leading to an increase in milling force [104]. Cutting forces (F_X and F_Y) are mostly affected by feed rate (~31%) and UP. Minimum cutting force can be achieved at a spindle speed of 2625 rpm, 100% UP, a depth of cut of 0.2 mm, and a feed rate of 10 mm/min. Surfaces produced by UAM exhibit better frictional stability, characterized by significantly reduced fluctuations in the friction coefficient curve. Compared to conventionally milled surfaces, UAM surfaces exhibit approximately a 20% reduction in friction coefficient and a 140% increase in oil film bearing capacity under optimized machining parameters [105], resulting in more uniform surfaces with fewer defects, as shown in **Figure 14(c)** [103].

5.4.2 Laser-assisted milling (LAMill)

Laser-assisted milling (LAMill) is a thermally assisted machining technique, in which a laser heat source preheats the workpiece ahead of the cutting tool, locally softening the material to facilitate improved machinability [111]. Under the combined effect of the milling tool and the laser heating source, three areas are produced: the milling-affected area, the laser-affected area, and the combined area (laser and milling) [112]. These areas can be differentiated through a schematic diagram, as shown in **Figure 15**.

LAMill reduces the cutting force and tool wear rate in both cases of high or low MRRs [113]. Better surface finish, along with improved surface characteristics, is another potential benefit of this technique. It is reported that a 7–40% improvement can be achieved in surface quality while employing LAMill [114]. Ahuir-Torres et al. [115] employed a picosecond laser on AA 2024-T3, enhancing surface wettability and corrosion resistance.

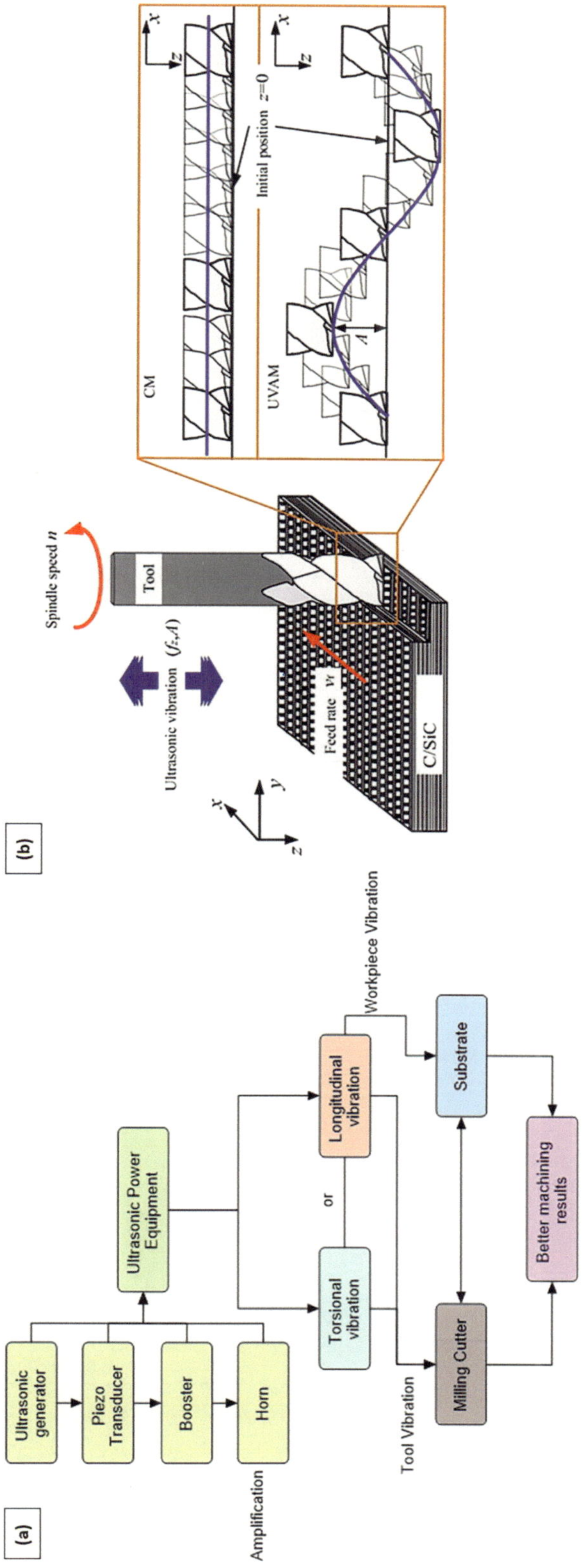

Figure 13.
UAM: (a) process flowchart and (b) schematic diagram [106].

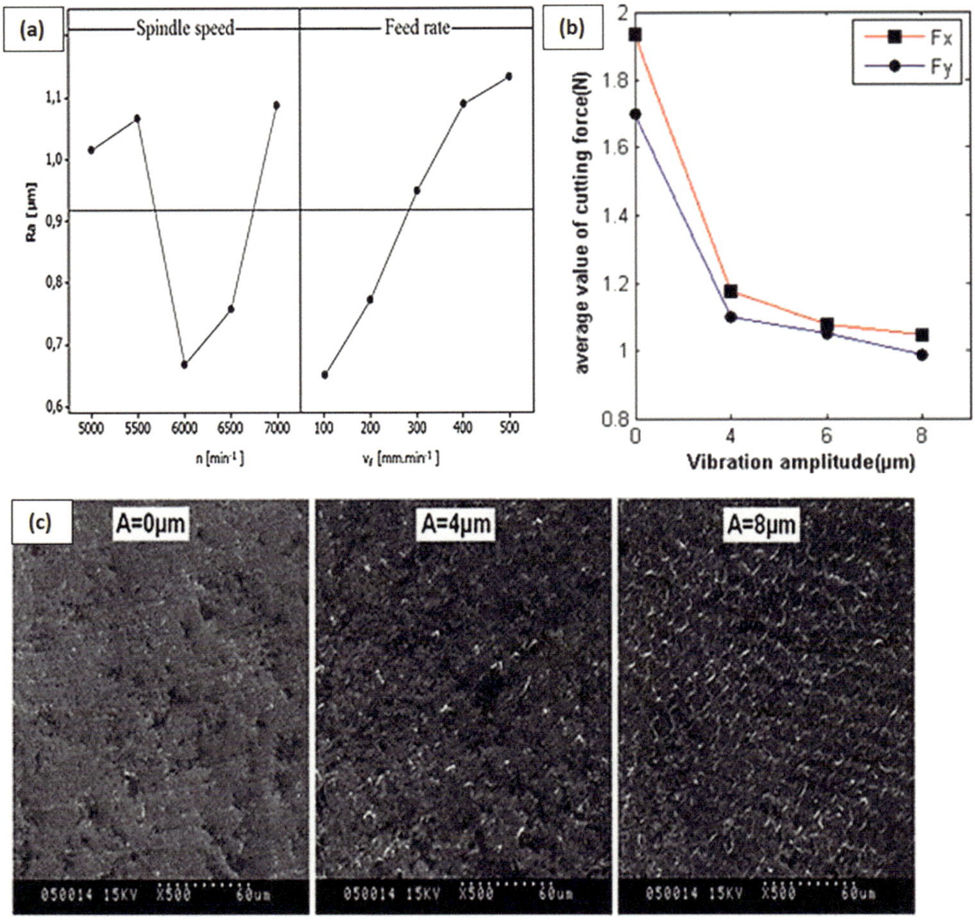

Figure 14.
Effect of UAM parameters: (a) surface roughness vs. spindle speed and feed rate [109], (b) cutting force vs. vibration amplitudes [103], and (c) SEM of finished surface with UAM [103].

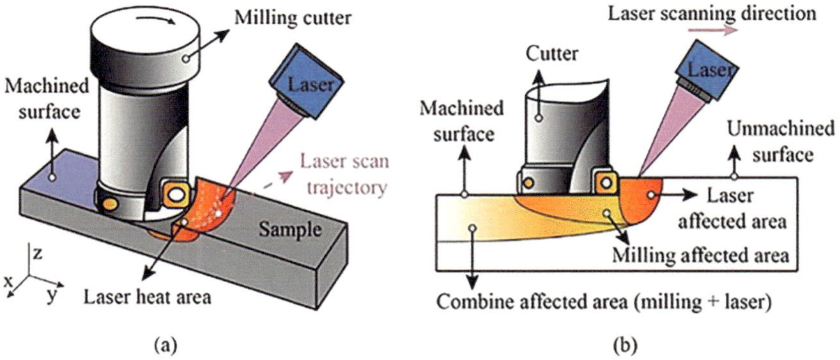

Figure 15.
Schematic of LAMill: (a) scan trajectory for uniform heating and (b) heat-influenced layer [112].

LAMill is mainly used to deal with difficult-to-cut materials such as ceramics, hard alloys, and steels [116, 117]. The important parameters of LAMill, in addition to cutting parameters, include the laser pulse energy and frequency, average power, track displacement, scanning path and its trajectory, and overlapping of the pulse and laser scan. The highest pulse energy, increasing pulse duration, and average power generate maximum MRR. ANOVA of Yb fiber laser milling of aluminum 6082 shows that surface roughness is significantly affected by track displacement (60%), pulse overlap (25%), and scanning strategy (14%), whereas MRR is only influenced by track displacement. This flexibility in pulse overlap and scanning strategy enables improvements in surface roughness without significantly affecting the MRR. Williams et al. [118] also reported that an increase in frequency increases the overlapping of subsequent pulses, resulting in increased debris and surface roughness, as shown in **Figure 16**(**a** and **b**). Campanelli et al. [119] preferred a parallel hatching scanning strategy over a mixed one and recommended a range of overlap (40%–70%) in order to achieve low surface roughness up to 5 μm, as shown in **Figure 16(c)** [119]. Lamp current intensity is another factor affecting machining performance. It exerts a stronger influence on micro-channel dimensions in aluminum alloys than pulse frequency or scan speed. It is difficult to achieve a micro-channel size of 100 μm for aluminum alloy due to its low melting point. A range of micro-channel sizes (125–150 μm) through Nd:YAG pulsed laser can be achieved for AA-2024 with optimized parameters at 83.05% lamp intensity, 30 kHz pulse frequency, and 300 mm/s scan speed [120].

5.4.3 Induction-assisted milling (IAM)

Induction-assisted milling is not only suitable for aluminum alloys but can also be employed on other difficult-to-machine materials, such as titanium gamma alloys [121]. IAM is comparatively cheaper than LAMill due to its low equipment cost [122]. A coil is used to preheat the work sample ahead of the milling cutter. Different schematics of IAM are shown in **Figure 17**. Usually, induction heating is applied to the substrate; however, Luo et al. [123] employed induction heating on the tool instead of the workpiece. This option is more suitable if the substrate material has a low melting point or if thermal effects on the surface area of the work part need to be avoided. Ha and Lee [124] performed a comparative study on the use of single-source and multisource assisted milling. They utilized LAMill, IAM, and a combination of both (laser and induction). The machining mechanism in IAM follows the same basic

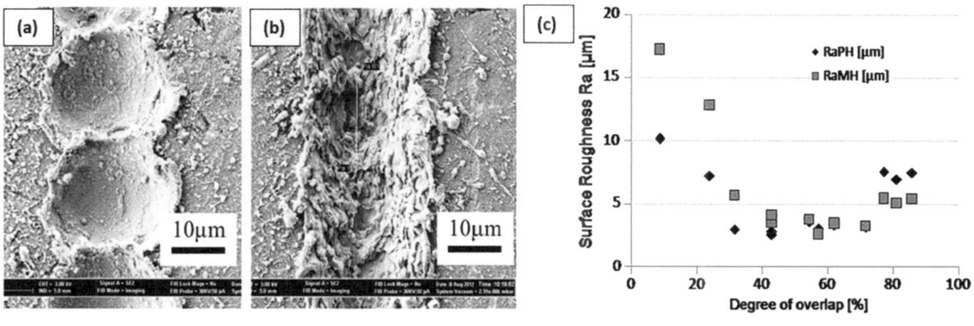

Figure 16.
Effect of pulse and scan overlap on surface roughness: (a) 30 kHz, (b) 60 kHz [118], and (c) scanning strategy [119].

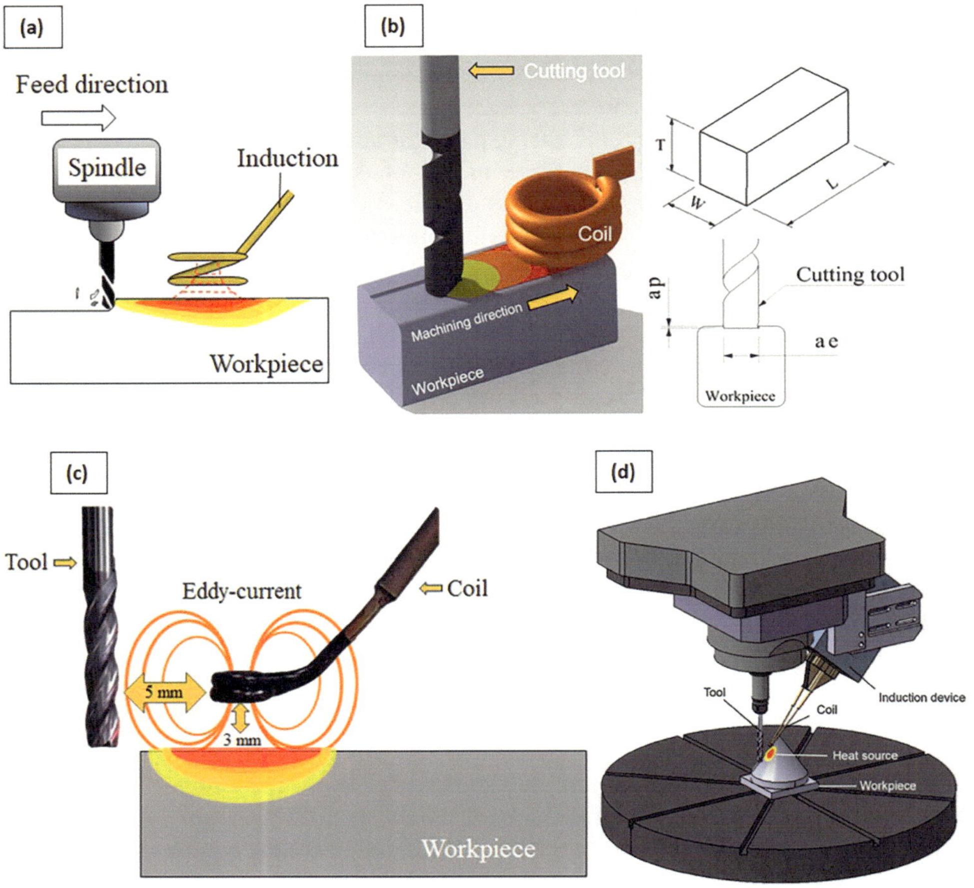

Figure 17.
Schematics of IAM: (a–b) different components of IAM, (c) distances among the tool, coil, and workpiece, and (d) IAM to mill a conical surface.

principle as thermal-assisted milling (TAM). The only differences are the source of heat, the ease of use, and the cost of the heating setup. The advantage of IAM lies in the flexibility of the heated coil, which can be directed at desired locations on the workpiece, especially when dealing with contoured or inclined surfaces, as depicted in **Figure 17(d)**. However, the limitation of this technique is its inadequacy for complex and deep milling features. It is more suitable for face milling, shoulder milling, end milling, and other milling operations where the milling cutter is subjected to wider areas of the substrate, and the placement of the induction coil does not pose a challenge. Another reason for this limitation is that the placement of the induction coil must maintain a reasonable distance from the milling cutter and workpiece surfaces, for example, 5 mm from the tool and 3 mm from the workpiece surface, as shown in **Figure 17(b)** [125].

5.4.4 PAM

Plasma-assisted milling is another process under the category of external energy-assisted machining. The schematic of PAM, extracted from [126], and the

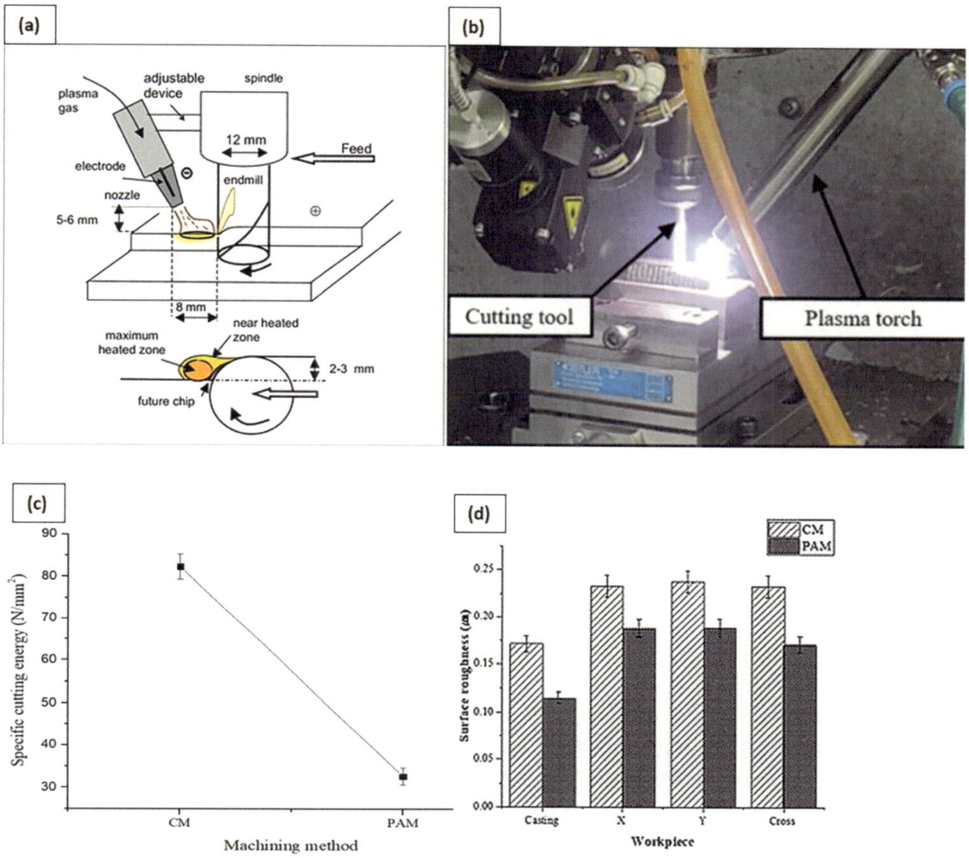

Figure 18.
PAM: (a) schematic diagram [126], (b) milling setup [127], (c) cutting energy comparison [129], and (d) surface roughness comparison [127].

actual setup, extracted from [127], are shown in **Figure 18**(**a** and **b**). The process mechanism of PAM follows the fundamental concept of heating the substrate to assist the milling cutter in order to reduce cutting forces. PAM enhances cutting performance and tool life by thermally softening the workpiece ahead of the cutting tool, eliminating the need for cutting oil or coolant [128]. Through the use of PAM, the cutting energy can be significantly reduced [129]. Moreover, in comparison with conventional milling, the surface roughness is also reduced through PAM [127].

PAM has the potential to significantly reduce the specific cutting energy compared to the conventional milling process. More than a 60% reduction in cutting energy and a 70% reduction in surface roughness of plasma-assisted milled titanium alloys are reported in [129]. Employing PAM for the steel alloy (SS 304), cutting forces and surface roughness can be reduced by 29% and 32%, respectively [130]. Compared with conventional milling of Inconel 718, PAM allows for a reduction in cutting forces and surface roughness by 20%. Additionally, this technique significantly reduces burr formation [127]. As detailed in [131], nanostructured aluminum nitride (AlN) powder can be synthesized via mechanical milling of a mixture of aluminum and lithium hydroxide monohydrate, using both conventional and discharge PAM techniques. AlN powders can be synthesized within one hour of milling using either traditional or PAM techniques.

Notably, PAM has not been directly found in the literature to perform milling of aluminum alloys. However, based on its significant impact on milling performance measures in the case of other materials, PAM has the potential to overcome challenges in the milling of hard aluminum alloys.

6. Conclusions

Given the extensive industrial applications of aluminum, the machining of aluminum and its alloys remains a key area of interest for both machinists and researchers. Compared to well-known hard materials such as titanium, steel, and super alloys, machining of aluminum is relatively easy. However, material adhesion, tool wear, dimensional variation, burr formation, and residual stresses are common problems reported during the milling of aluminum alloys due to their soft nature. The milling of aluminum and its alloys is influenced by alloying composition, cutting tool material and its geometry, cutting parameters, and cooling environment.

Based on insights from various researchers, the following conclusions can be drawn to achieve better machining results:

- Tangential and radial forces have a significant effect on cutting forces, which can be reduced by using high cutting speed, low feed rate, and depth of cut. The most suitable cutting tools can be either diamond-coated or diamond cutting tools due to their low affinity for aluminum.

- In aluminum alloys, residual stresses are mostly compressive and become tensile with an increase in cutting temperature, cutting speed, feed rate, and flank wear.

- Aluminum is a soft material with high thermal conductivity, which makes thermal expansion and dimensional variation difficult to control. An increase in TER, feed rate, cutting speed, and a low depth of cut, combined with an adequate coolant, reduces the overall cutting temperature, ultimately lowering the overall stress level.

- Burr formation in the milling of aluminum alloys is very common. Burr formation is directly influenced by the increase in tool wear, feed rate, and friction between the workpiece and tool contact.

- Uncoated tools are more effective in achieving fine surface finishes at cutting speeds up to 30 m/min, whereas coated tools excel at higher speeds (50–60 m/min). Moreover, minimizing the cutting-edge radius and feed rate, while maximizing spindle speed, significantly reduces surface roughness.

- Tool wear increases with spindle speed and depth of cut, and decreases with an increase in feed rate. Flank wear is commonly observed during the milling of aluminum alloys, whereas notch wear is more prominent in harder aluminum alloys such as AA-2024, A-390, and AA-7075.

- Mostly, pretreatments and coatings do not weaken the cutting tool and help in reducing cutting forces, material adhesion, and burr formation. DLC and diamond-coated tools exhibit exceptionally low friction coefficients and minimal chemical affinity toward aluminum alloys, making them highly suitable for use with aluminum alloys.

- To mitigate premature failure in high-speed cutters, tool materials should possess excellent impact resistance, wear resistance, and thermal fatigue strength – ideally based on ultrafine grain carbide or a fine-grain matrix – combined with a high-performance wear-resistant coating.

- Moderate vibrations induced either in the workpiece or in the cutting tool reduce the interaction time between the tool and the work material, thereby significantly reducing cutting forces and cutting temperature. Tool life and surface integrity are consequently improved.

- Nanoparticles, up to a 0.5% concentration in the MQL technique, can improve various machining characteristics, including reducing cutting temperature, cutting forces, burr formation, flank wear, surface roughness, and chip adhesion.

- EEAM approaches (such as LAM, VAM, IAM, and PAM) have great potential to address challenges associated with the conventional milling of aluminum alloys. However, these approaches are relatively less utilized in milling aluminum alloys compared to other materials.

Acknowledgment

The authors would like to thank Al Yamamah University for supporting this work.

Author details

Naveed Ahmed*, Osamah Abdulhameed, Madiha Rafaqat and Wasim Alshammary
Department of Industrial Engineering, College of Engineering, Al Yamamah University, Riyadh, Saudi Arabia

*Address all correspondence to: n_ahmed@yu.edu.sa

References

[1] Songmene V, Khettabi R, Zaghbani I, Kouam J, Djebar A. Machining and machinability of aluminum alloys. Aluminium Alloys: Theory and Applications. 2011. DOI: 10.5772/14888

[2] Kalpakjian S, Schmid SR. Manufacturing Engineering and Technology. 6th ed. United Kingdom: Pearson; 2009

[3] Edward M. Trent, Paul K. Wright. Management of technology. In: Edward M. Trent, Paul K (Editor). Wright, Metal Cutting (Fourth Edition). Butterworth-Heinemann; 2000. p. 411–423. DOI: 10.1016/B978-075067069-2/50015-2

[4] Degarmo EP, Black JT, and Kohser RA. Nonferrous Metals and Alloys. In Materials and Process in Manufacturing. 9th ed. Hoboken: Wiley; 2003. p. ch.7

[5] Davis JR. Aluminum and aluminum alloys. In Alloying: Understanding the Basics. Netherlands: ASM International;2001. pp. 351–416. DOI: 10.1361/autb2001p351

[6] The Aluminum Association, "Aluminum: The element of sustainability," 2011

[7] Sreejith PS. Machining of 6061 aluminium alloy with MQL, dry and flooded lubricant conditions. Materials Letters. 2008;**62**:276–278. DOI: 10.1016/j.matlet.2007.05.019

[8] Cock T. Aluminium—a Light Metal: A Brief Introduction to Aluminium. Oslo: TALAT (Training in Aluminium Application Technologies); 1999

[9] Norberto López de Lacalle L, Campa FJ, Lamikiz A. Milling. In: Paulo Davim, J. Modern Machining Technology. Woodhead Publishing; 2011. p. 213–303. DOI: 10.1533/9780857094940.213

[10] Chang H, Li S, Shi R. Design and manufacturing technology of high speed milling cutter for aluminum alloy. Procedia Engineering. 2017;**174**: 630–637. DOI: 10.1016/j.proeng.2017.01.200

[11] Starke EA, Rashed HMMA. Alloys: Aluminum. Reference Module in Materials Science and Materials Engineering. 2016;**2017**. DOI: 10.1016/B978-0-12-803581-8.09210-9

[12] D H. of the Army, "US Army - Fundamentals of Machine Tools TC 9-524,". Washington D.C: Headquarters Department of the Army; 1996. 316 p.

[13] Sandvik C. Metal Cutting Technical Guide. Sandviken, Sweden: Sandvik Inc; 2011

[14] Nghiep TN, Sarhan AAD, Aoyama H. Analysis of tool deflection errors in precision CNC end milling of aerospace aluminum 6061-T6 alloy. Measurement: Journal of the International Measurement Confederation. 2018;**125**:476–495. DOI: 10.1016/j.measurement.2018.05.011

[15] Thamban I, Abraham BC, Kurian S. Machining characteristics analysis of 6061-T6 aluminium alloy with diamond coated and uncoated tungsten carbide tool. International Journal of Latest Research in Science and Technology. 2013;**2**(1):553–557

[16] Luo M, Chong Z, Liu D. 2018. Cutting forces measurement for milling process by using working tables with

integrated PVDF thin-film sensors. Sensors. **18**(11):4031. DOI: 10.3390/s18114031

[17] Salguero J, Batista M, Calamaz M, Girot F, Marcos M. Cutting forces parametric model for the dry high speed contour milling of aerospace aluminium alloys. Procedia Engineering. 2013;**63**:735–742. DOI: 10.1016/j.proeng.2013.08.215

[18] Dikshit MK, Puri AB, Maity A, Banarjee AJ. "Determining Cutting Force Coefficient from Instantaneous Cutting Force in Ball-End Milling,". AIMTDR. 2014;1–7. DOI: 10.1038/nmat4618

[19] Campatelli G, Scippa A. "Prediction of milling cutting force coefficients for Aluminum 6082-,". Procedia CIRP. 2012;**1**:563–568. DOI: 10.1016/j.procir.2012.04.100

[20] Lmalghan R, C KRM, Arunkumar S, Rao SS, Herbert MA. 2018. "Machining parameters optimization of AA6061 using response surface methodology and particle swarm optimization,". The International Journal of Advanced Manufacturing Technology. **19**(5):695–704. DOI: 10.1007/s12541-018-0083-2

[21] Kumar M, Baran A, Maity A. 2014. Experimental study of cutting forces in ball end milling of Al2014-T6 using response surface methodology. MSPRO. **6**(Icmpc):612–622. DOI: 10.1016/j.mspro.2014.07.076

[22] Subramanian M, Sakthivel M, Sooryaprakash K, Sudhakaran R. Optimization of cutting parameters for cutting force in shoulder milling of Al7075-T6 using response surface methodology and genetic algorithm. Procedia Engineering. 2013;**64**:690–700. DOI: 10.1016/j.proeng.2013.09.144

[23] Niu Z, Jiao F, Cheng K. An innovative investigation on chip formation mechanisms in micro-milling using natural diamond and tungsten carbide tools. Journal of Manufacturing Processes. 2018;**31**:382–394. DOI: 10.1016/j.jmapro.2017.11.023

[24] Torres CD, Heaney PJ, Sumant AV, Hamilton MA, Carpick RW, Pfefferkorn FE. 2009. Analyzing the performance of diamond-coated micro end mills. International Journal of Machine Tools and Manufacture. **49**(7–8):599–612. DOI: 10.1016/j.ijmachtools.2009.02.001

[25] Ng E-G, Szablewski D, Dumitrescu M, Elbestawi MA, Sokolowski JH. "High speed face milling of aluminium silicon alloy casting,". CIRP Annals. 2004;**53**:69–72

[26] Fu W, Cohen PH, Ruud CO. 2009. Experimental investigation of the machining induced residual stress tensor under mechanical loading. Journal of Manufacturing Processes. **11**(2):88–96. DOI: 10.1016/j.jmapro.2009.11.001

[27] Rimpault X, Il A, Chatelaina J-F, Lalondec J-F, Balazinskib M. 2018. "Workpiece subsurface temperature study during aluminum skin milling in slotting and ramping,". Procedia CIRP. **77**(Hpc):417–420. DOI: 10.1016/j.procir.2018.08.295

[28] Li Y, Gan W, Zhou W, Li D. 2023. Review on residual stress and its effects on manufacturing of aluminium alloy structural panels with typical multi-processes. Chinese Journal of Aeronautics. **36**(5):96–124. DOI: 10.1016/j.cja.2022.07.020

[29] Tang ZT, Liu ZQ, Wan Y, Ai X. Study on residual stresses in milling aluminium alloy 7050-T7451. In Yan X-T, Jiang C, Eynard B, editors. Advanced Design and Manufacture to Gain a Competitive Edge. London: Springer;2008. pp. 169–178. DOI: 10.1007/978-1-84800-241-8_18

[30] Xu Y, Yue C, Chen Z, Li M, Wang L, Liu X. 2023. "Finite element simulation of residual stress in milling of aluminum alloy with different passes,". The International Journal of Advanced Manufacturing Technology. **127**(9):4199–4210. DOI: 10.1007/s00170-023-11795-2

[31] Yi S, Wu Y, Gong H, Peng C, He Y. 2021. Experimental analysis and prediction model of milling-induced residual stress of aeronautical aluminum alloys. Applied Sciences. **11**(13):5881. DOI: 10.3390/app11135881

[32] Denkena B, De Leon L. 2008. Milling induced residual stresses in structural parts out of forged aluminium alloys. Journal of Machining and Machinability of .... **4**(4):2–5. DOI: 10.1504/IJMMM.2008.023717

[33] Kuczmaszewski J, Łogin W, Pieśko P, Zagórski I, "State of Residual Stresses After the Process of Milling Selected Aluminium Alloys", [Online]. Available: https://www.astrj.com/State-of-Residual-Stresses-After-the-Process-of-Milling-Selected-Aluminium-Alloys,81768,0,2.html. [Accessed: 2025-September-01]

[34] Jomaa W, Songmene V, Bocher P. 2014. Surface finish and residual stresses induced by orthogonal dry machining of AA7075-T651. Materials. **7**(3):1603–1624. DOI: 10.3390/ma7031603

[35] Parez I, *et al.* Effect of cutting speed on the surface integrity of face milled 7050-T7451 aluminium workpieces. Procedia CIRP. 2018;**71**:460–465

[36] Fuh KH, Wu CF. 1995. A residual-stress model for the milling of aluminum alloy (2014-T6). Journal of Materials Processing Tech. **51**(1–4):87–105. DOI: 10.1016/0924-0136(94)01355-5

[37] Tang ZT, Liu ZQ, Pan YZ, Wan Y, Ai X. 2009. The influence of tool flank wear on residual stresses induced by milling aluminum alloy. Journal of Materials Processing Technology. **209** (9):4502–4508. DOI: 10.1016/j.jmatprotec.2008.10.034

[38] Jiao L, Wang X, Qian Y, Liang Z, Liu Z. 2015. Modelling and analysis for the temperature field of the machined surface in the face milling of aluminium alloy. International Journal of Advanced Manufacturing Technology. **81** (9–12):1797–1808. DOI: 10.1007/s00170-015-7231-y

[39] Yang K, Liang YC, Zheng KN, Bai QS, Chen WQ. 2011. Tool edge radius effect on cutting temperature in micro-end-milling process. International Journal of Advanced Manufacturing Technology. **52**(9–12):905–912. DOI: 10.1007/s00170-010-2795-z

[40] Kaushik VS, Subramanian M, Sakthivel M. 2018. Optimization of processes parameters on temperature rise in CNC end milling of Al 7068 using hybrid techniques. Materials Today: Proceedings. 5(2, Part 2):7037–7046. DOI: 10.1016/j.matpr.2017.11.367

[41] Rimpault X, Il A, Chatelain J-F, Lalonde J-F, Balazinski M. Workpiece subsurface temperature study during aluminum skin milling in slotting and ramping. Procedia CIRP. 2018;77: 417–420. DOI: 10.1016/j.procir.2018.08.295

[42] Bhirud NL, Gawande RR. 2017. Optimization of process parameters during end milling and prediction of work piece temperature rise. Archive of Mechanical Engineering. **64**(3):327–346. DOI: 10.1515/meceng-2017-0020

[43] Denkena B, Brüning J, Niederwestberg D, Grabowski R. Influence of machining parameters on heat generation during milling of aluminum alloys. Procedia CIRP. 2016;**46**:39–42. DOI: 10.1016/j.procir.2016.03.192

[44] Santos Jr MC, Machado AR, Sales WF. Machining of aluminum alloys : A review. The International Journal of Advanced Manufacturing Technology. 2016;**86**:3067–3080. DOI: 10.1007/s00170-016-8431-9

[45] Vázquez E, Rodríguez CA, Elías-zúñiga A. An experimental analysis of process parameters to manufacture metallic micro-channels by micro-milling. The International Journal of Advanced Manufacturing Technology. 2010;**51**:945–955. DOI: 10.1007/s00170-010-2685-4

[46] Albertí M, Ciurana J, Rodriguez CA. 2007. Experimental analysis of dimensional error vs. cycle time in high-speed milling of aluminium alloy. International Journal of Machine Tools and Manufacture. **47**(2):236–246. DOI: 10.1016/j.ijmachtools.2006.04.009

[47] Zawada-Michałowska M, Pieśko P, Józwik J, Legutko S, Kukiełka L. 2021. A comparison of the geometrical accuracy of thin-walled elements made of different aluminum alloys. Materials. **14**(23):7242. DOI: 10.3390/ma14237242

[48] Zawada-Michałowska M, Kuczmaszewski J, Pieśko P. 2022. Effect of the geometry of thin-walled aluminium alloy elements on their deformations after milling. Materials. **15**(24):9049. DOI: 10.3390/ma15249049

[49] Shinge AR, Dabade UA. The effect of process parameters on material removal rate and dimensional variation of channel width in micro-milling of aluminium alloy 6063 T6. Procedia Manufacturing. 2018;**20**:168–173. DOI: 10.1016/j.promfg.2018.02.024

[50] Kiswanto G, Zariatin DL, Ko TJ. The effect of spindle speed, feed-rate and machining time to the surface roughness and burr formation of aluminum alloy 1100 in micro-milling operation. Journal of Manufacturing Processes. 2014;**16**:435–450. DOI: 10.1016/j.jmapro.2014.05.003

[51] Niknam SA, Songmene V. 2013. Simultaneous optimization of burrs size and surface finish when milling 6061-T6 aluminium alloy. International Journal Of Precision Engineering And Manufacturing. **14**(8):1311–1320. DOI: 10.1007/s12541-013-0178-8

[52] Haddag B, Atlati S, Nouari M, Moufki A. Dry machining aeronautical aluminum alloy AA2024-T351: Analysis of cutting forces, chip segmentation and built-up edge formation. Metals. 2016; **6**(9):9. DOI: 10.3390/met6090197

[53] Gómez-Parra A, Álvarez-Alcón M, Salguero J, Batista M, Marcos M. 2013. Analysis of the evolution of the built-up edge and built-up layer formation mechanisms in the dry turning of aeronautical aluminium alloys. Wear. **302**(1):1209–1218. DOI: 10.1016/j.wear.2012.12.001

[54] Kirubadurai B, Kanagaraja K, Jegadeeswari G. 2022. Impact of BUE

on surface ruggedness for machining of aluminium matrix composites (6061/fly ash). Advances in Materials and Processing Technologies. **8** (1):542–549. DOI: 10.1080/2374068X.2020.1823656

[55] Dasch JM, *et al.* The effect of free-machining elements on dry machining of B319 aluminum alloy. Journal of Materials Processing Technology. 2009;**209**(10):4638–4644. DOI: 10.1016/j.jmatprotec.2008.11.041

[56] Kelly JF, Cotterell MG. Minimal lubrication machining of aluminium alloys. Journal of Materials Processing Technology. 2002;**120**:327–334

[57] Edem IF, Balogun VA. Sustainability analyses of cutting edge radius on specific cutting energy and surface finish in side milling processes. The International Journal of Advanced Manufacturing Technology. 2018;**95**:3381–3391

[58] Niknam SA, Songmene V. Analytical modelling of slot milling exit burr size. The International Journal of Advanced Manufacturing Technology. 2014;**73**:421–432. DOI: 10.1007/s00170-014-5758-y

[59] Niknam SA, Songmene V. Analysis of friction and burr formation in slot milling. Procedia CIRP. 2014;**17**:755–759. DOI: 10.1016/j.procir.2014.02.032

[60] Wang Y, Li L, Lingling L, Cai W. Exploring the effect of un-deformed chip parameters on energy consumption for energy efficiency improvement in the milling. Procedia CIRP. 2018;**72**:1380–1385. DOI: 10.1016/j.procir.2018.03.075

[61] Alharthi NH, Bingol S, Abbas AT, Ragab AE, El-Danaf EA, Alharbi HF. Optimizing cutting conditions and prediction of surface roughness in face milling of AZ61 using regression analysis and artificial neural network. Advances in Materials Science and Engineering. 2017;**2017**:1–8. DOI: 10.1155/2017/7560468

[62] Tomadi SH, Ghani JA, Haron CHC, Ayu HM, Daud R. Effect of cutting parameters on surface roughness in end milling of AlSi/AlN metal matrix composite. Procedia Engineering. 2017;**184**:58–69. DOI: 10.1016/j.proeng.2017.04.071

[63] Jing X, *et al.* An investigation of surface roughness in micro-end-milling of metals. Australian Journal of Mechanical Engineering. 2017;**15** (3):166–174. DOI: 10.1080/14484846.2016.1211472

[64] Yahya E, Ding G, Qin S. 2016. Prediction of cutting force and surface roughness using Taguchi technique for aluminum alloy AA6061. Australian Journal of Mechanical Engineering. **14** (3):151–160. DOI: 10.1080/14484846.2015.1093220

[65] Ciftci I, Turker M, Seker U. 2004. Evaluation of tool wear when machining SiCp-reinforced Al-2014 alloy matrix composites. Materials & Design. **25** (3):251–255. DOI: 10.1016/j.matdes.2003.09.019

[66] Nurhaniza M, Ariffin MKAM, Mustapha F, Baharudin BTHT. Analyzing the effect of machining parameters setting to the surface roughness during end milling of CFRP-aluminium composite laminates. International Journal of Manufacturing Engineering. 2016;**2016**:1–9

[67] Jayakumar K, “Study of cutting force and surface roughness in Ball nose

end milling of vacuum hot pressed A356 alloy/SiC p metal matrix composite," *Materials Today: Proceedings*, vol. **5**, pp. 6526–6533, 2018, DOI: 10.1016/j.matpr.2017.11.307

[68] Wang SJ, To S, Cheung CF. 2012. Effect of workpiece material on surface roughness in ultraprecision raster milling. Materials and Manufacturing Processes. **27**(10):1022–1028. DOI: 10.1080/10426914.2011.654165

[69] Cardoso P, Davim JP. 2010. Optimization of surface roughness in micromilling. Materials and Manufacturing Processes. **25**(10): 1115–1119. DOI: 10.1080/10426914.2010.481002

[70] Unnikrishna J, Sanghrajka I, Shunmugavel M, Muthuramalingam T, Goldberg M, Littlefair G. 2018. Optimisation of multiple response characteristics on end milling of aluminium alloy using Taguchi-Grey relational approach. Measurement. **124** (April):291–298. DOI: 10.1016/j.measurement.2018.04.052

[71] Muñoz-Escalona P, Shokrani A, Newman ST. 2016. Optimal cutting conditions towards sustainable machining when slot milling aluminium alloy. Advances in Materials and Processing Technologies. **2**(4):480–489. DOI: 10.1080/2374068X.2016.1247228

[72] Fukui H, Okida J, Omori N, Moriguchi H, Tsuda K. 2004. Cutting performance of DLC coated tools in dry machining aluminum alloys. Surface and Coatings Technology. **187**(1):70–76. DOI: 10.1016/j.surfcoat.2004.01.014

[73] Kannan IS, Ghosh A. Dry machining of AA7075 by H-DLC coated carbide end mill. Procedia Materials Science. 2014;**5**:2615–2621. DOI: 10.1016/j.mspro.2014.07.522

[74] Roy P, Sarangi SK, Ghosh A, Chattopadhyay AK. 2009. Machinability study of pure aluminium and Al – 12 % Si alloys against uncoated and coated carbide inserts. International Journal of Refractory Metals & Hard Materials. **27**(3):535–544. DOI: 10.1016/j.ijrmhm.2008.04.008

[75] KuczmAszewsKi J, PieśKo P. Wear of milling cutters resulting from high silicon aluminium alloy cast AlSi21CuNi machining. Maintenance and Reliability. 2014;**16**(1):37–41

[76] Xu K, Zou B, Wang Y, Guo P, Huang C, Wang J. An experimental investigation of micro-machinability of aluminum alloy 2024 using Ti (C7N3)-based cermet micro end-mill tools. Journal of Materials Processing Technology. 2016;**235**:13–27. DOI: 10.1016/j.jmatprotec.2016.04.011

[77] Kasim MS, Atan MH, Haron CHC, Ghani JA. Analysis of tool performance during ball-end milling of aluminium alloy 6061-T6. Applied Mechanics and Materials. 2015;**761**:318–323. DOI: 10.4028/www.scientific.net/AMM.761.318

[78] Calatoru VD, Balazinski M, Mayer JRR, Paris H, L'Espérance G. 2008. Diffusion wear mechanism during high-speed machining of 7475-T7351 aluminum alloy with carbide end mills. Wear. **265**(11–12):1793–1800. DOI: 10.1016/j.wear.2008.04.052

[79] Sugihara T, Singh P, Enomoto T. Development of novel cutting tools with dimple textured surfaces for dry machining of aluminum alloys. Procedia Manufacturing. 2017;**14**:

111–117. DOI: 10.1016/j.promfg.2017.11.013

[80] Yan J, Murakami Y, and Davim JP. Tool Design, Tool Wear and Tool Life. In Machining Dynamics. 1st. Cheng K editor. London: Springer; 2002. ch. 5. pp. 117–149. DOI: 10.1007/978-1-84628-368-0_5

[81] Lahres M, Muller-Hummel P, Doerfel O. Applicability of different hard coatings in dry milling aluminium alloys. Surface and Coatings Technology. 1997;**91**:116–121

[82] Kannan S, Ghosh A. 2018. Anti-friction and wetting behavior of a new polymer composite coating towards aluminium and dry machining of AA2024 alloy by coated end mills. Journal of Materials Processing Technology. **252**(July 2017):280–293. DOI: 10.1016/j.jmatprotec.2017.09.033

[83] Tsao CC. Grey–Taguchi method to optimize the milling parameters of aluminum alloy. The International Journal of Advanced Manufacturing Technology. 2007;**40**:41–48. DOI: 10.1007/s00170-007-1314-3

[84] Kuram E, Ozcelik B. 2013. Multi-objective optimization using Taguchi based grey relational analysis for micro-milling of Al 7075 material with ball nose end mill. Measurement: Journal of the International Measurement Confederation. **46**(6):1849–1864. DOI: 10.1016/j.measurement.2013.02.002

[85] Dixit US, Sarma AK, Davim JP. Machining with minimal cutting fluid. In: Environmentally Friendly Machining. Boston, MA: Springer; 2012. p. 9–18. DOI: 10.1007/978-1-4614-2308-9

[86] Astakhov V. Tribology of Metal Cutting. 1st ed. Netherlands: Elsevier Science; 2006

[87] Komanduri JDR. Tool materials encyclopedia of chemical technology. Carbide Tool Journal. 1983;(5):273–309

[88] S. P, Singh JSDT.A review of near dry machining/minimum quantity lubrication machining of difficult to machine alloys. International Journal of Machining and Machinability of Materials. 2016;**18**(3). DOI: 10.1504/IJMMM.2016.076276

[89] Pervaiz S, Anwar S, Qureshi I, Ahmed N. Recent advances in the machining of titanium alloys using Minimum Quantity Lubrication (MQL) based techniques. International Journal of Precision Engineering and Manufacturing-Green Technology. 2017. DOI: 10.1007/s40684-019-00033-4

[90] Schey JA. Tribology in Metalworking: Friction, Lubrication and Wear. Ohio: American Society for Metals, Metal Parks; 1983

[91] Lacalle D, Angulo C, Lamikiz A. Experimental and numerical investigation of the effect of spray cutting fluids in high speed milling. Journal of Materials Processing Technology. 2006;**172**:11–15. DOI: 10.1016/j.jmatprotec.2005.08.014

[92] Tosun N, Pihtili H. Gray relational analysis of performance characteristics in MQL milling of 7075 Al alloy. The International Journal of Advanced Manufacturing Technology. 2010;**46**:509–515. DOI: 10.1007/s00170-009-2118-4

[93] Tosun N, Huseyinoglu M. Effect of MQL on surface roughness in milling of AA7075-T6. Materials and Manufacturing Processes. 2010;**25**: 793–798. DOI: 10.1080/10426910903496821

[94] Kishawy HA, Dumitrescu M, Ng E, Elbestawi MA. Effect of coolant strategy on tool performance, chip morphology and surface quality during high-speed machining of A356 aluminum alloy. International Journal of Machine Tools and Manufacture. 2005;**45**:219–227. DOI: 10.1016/j.ijmachtools.2004.07.003

[95] Khettabi R, Nouioua M, Djebara A, Songmene V. Effect of MQL and dry processes on the particle emission and part quality during milling of aluminum alloys. The International Journal of Advanced Manufacturing Technology. 2017;**92**:2593–2598. DOI: 10.1007/s00170-017-0339-5

[96] Imthiyas A, Jayasuriya D, Mani MJ. Experimental investigation of milling operation on aluminum alloy 6063 by MQL technique. International Journal of Chemical Sciences. 2016;**14**(4): 3113–3118

[97] Sayuti M, Sarhan AAD, Hamdi M. An investigation of optimum SiO2 nanolubrication parameters in end milling of aerospace Al6061-T6 alloy. The International Journal of Advanced Manufacturing Technology. 2013;**67**:833–849. DOI: 10.1007/s00170-012-4527-z

[98] Rahmati B, Sarhan AAD, Sayuti M. Morphology of surface generated by end milling AL6061-T6 using molybdenum disulfide (MoS 2) nanolubrication in end milling machining. Journal of Cleaner Production. 2014;**66**:685–691. DOI: 10.1016/j.jclepro.2013.10.048

[99] Karim MSA, Derahman NA, Amran M, Isa MAM. Analysation of performances of CNC high speed milling machine using multi-walled carbon nanotubes as additives in cutting fluid. Materialwissenschaft And Werkstofftechnik. 2018;**49**: 494–499. DOI: 10.1002/mawe.201800028

[100] Najiha MS, Rahman MM, Kadirgama K. Performance of water-based TiO2 nanofluid during the minimum quantity lubrication machining of aluminium alloy, AA6061-T6. Journal of Cleaner Production. 2016;**135**:1623–1636. DOI: 10.1016/j.jclepro.2015.12.015

[101] Xu WX, Zhang LC. 2015. Ultrasonic vibration-assisted machining: Principle, design and application. Advances in Manufacturing. **3**(3):173–192. DOI: 10.1007/s40436-015-0115-4

[102] Kushan MC, Orak S, Uzunonat Y. Ultrasonic assisted machining methods : A review. International Journal of Advanced Engineering Research and Applications. 2017;**3**(September): 294–302

[103] Shen X, Zhang J, Li H, Wang J, Wang X. Ultrasonic vibration-assisted milling of aluminum alloy. The International Journal of Advanced Manufacturing Technology. 2012;**63**: 41–49. DOI: 10.1007/s00170-011-3882-5

[104] Shen X, Xu G. 2018. Study of milling force variation in ultrasonic vibration-assisted end milling. Materials and Manufacturing Processes. **33** (6):644–650. DOI: 10.1080/10426914.2017.1364846

[105] Xing D, Zhang J, Shen X, Zhao Y, Wang T. Tribological properties of ultrasonic vibration assisted milling aluminium alloy surfaces. Procedia - Social and Behavioral Sciences. 2013;**6**:539–544. DOI: 10.1016/j.procir.2013.03.008

[106] Liu Y, Liu Z, Wang X, Huang T. 2020. Experimental study on tool wear in ultrasonic vibration–assisted milling of C/SiC composites. The International Journal of Advanced Manufacturing Technology. **107**(1):425–436. DOI: 10.1007/s00170-020-05060-z

[107] Zarchi MMA, Razfar MR, Abdullah A. 2014. Investigation of the effect of cutting speed and vibration amplitude on cutting forces in ultrasonic-assisted milling. Proceedings of the Institution of Mechanical Engineers, Part B: Journal of Engineering Manufacture. **226**(7):1185–1191. DOI: 10.1177/0954405412439666

[108] Ko JH, Tan SW. 2013. Chatter marks reduction in meso-scale milling through ultrasonic vibration assistance parallel to tooling ' s axis. The International Journal of Advanced Manufacturing Technology. **14**(1):17–22. DOI: 10.1007/s12541-013-0003-4

[109] Marcel K, Jozef P. Investigation of ultrasonic assisted milling of aluminum alloy. Procedia Engineering. 2014;**69**:1048–1053. DOI: 10.1016/j.proeng.2014.03.089

[110] Chandra G, Mohan P. Machining forces in ultrasonic-vibration assisted end milling. Ultrasonics. 2018. DOI: 10.1016/j.ultras.2018.07.004

[111] Lee CM, Kim DH, Baek JT, Kim EJ. 2016. Laser assisted milling device: A review. International Journal of Precision Engineering and Manufacturing-Green Technology. **3**(2):199–208. DOI: 10.1007/s40684-016-0027-1

[112] Xu D, Liao Z, Axinte D, Sarasua JA, M'Saoubi R, Wretland A. Investigation of surface integrity in laser-assisted machining of nickel based superalloy. Materials & Design. 2020; **194**:108851. DOI: 10.1016/j.matdes.2020.108851

[113] Kumar M, Melkote SN. 2012. Process capability study of laser assisted micro milling of a hard-to-machine material. Journal of Manufacturing Processes. **14**(1):41–51. DOI: 10.1016/j.jmapro.2011.09.003

[114] Liu Q, Liu J, Ming Z, Cui B, Wang J. Research Progress on Laser-Assisted Precision Machining Technology. Micromachines. 2025; **16**(2):2. DOI: 10.3390/mi16020173

[115] Ahuir-Torres JI, Arenas MA, Perrie W, de Damborenea J. Influence of laser parameters in surface texturing of Ti6Al4V and AA2024-T3 alloys. Optics and Lasers in Engineering. 2018;**103**:100–109. DOI: 10.1016/j.optlaseng.2017.12.004

[116] Kovalenko V, *et al.* Laser milling of the intractable materials. In: Procedia CIRP. vol. **6**. Netherlands: Elsevier B.V; 2013. p. 504–509. DOI: 10.1016/j.procir.2013.03.068

[117] Chryssolouris G, Anifantis N, Karagiannis S. 1997. Laser assisted machining: An overview. Journal of Manufacturing Science and Engineering. **119**(4B):766. DOI: 10.1115/1.2836822

[118] Williams E, Brousseau EB, Rees A. 2014. Nanosecond Yb fibre laser milling of aluminium: Effect of process parameters on the achievable surface finish and machining efficiency. International Journal of Advanced Manufacturing Technology. **74**(5–8):769–780. DOI: 10.1007/s00170-014-6038-6

[119] Campanelli SL, Casalino G, Contuzzi N. Multi-objective

optimization of laser milling of 5754 aluminum alloy. Optics and Laser Technology. 2013;**52**:48–56. DOI: 10.1016/j.optlastec.2013.03.020

[120] Alahmari AM, Darwish S, Ahmed N. 2016. Laser Beam Micro-Milling (LBMM) of selected aerospace alloys. International Journal of Advanced Manufacturing Technology. **86** (9–12):2411–2431. DOI: 10.1007/s00170-015-8318-1

[121] Fan T, Yao C, Tan L, Cao Y, Sun Y, Tang W. The influence of induction-assisted milling on the machining characteristics and surface integrity of γ-TiAl alloys. Journal of Manufacturing Processes. 2024;**118**:215–227. DOI: 10.1016/j.jmapro.2024.03.012

[122] Bijanzad A, Munir T, Abdulhamid F. 2022. Heat-assisted machining of superalloys: A review. The International Journal of Advanced Manufacturing Technology. **118**(11):3531–3557. DOI: 10.1007/s00170-021-08059-2

[123] Luo J, Ding H, J. A, Shih. 2005. Induction-heated tool machining of elastomers—Part 2: Chip morphology, cutting forces, and machined surfaces. Machining Science and Technology. **9** (4):567–588. DOI: 10.1080/10910340500398233

[124] Ha J-H, Lee C-M. A study on the thermal effect by multi heat sources and machining characteristics of laser and induction assisted milling. Materials. 2019;**12**(7):7. 1032. DOI: 10.3390/ma12071032

[125] Kim EJ, Lee CM. A study on the optimal machining parameters of the induction assisted milling with Iinconel 718. Materials. 2019; **12**(2):2. DOI: 10.3390/ma12020233

[126] Lo´pez LN, de Lacalle JA, Sa´nchez AL, Celaya A. 2004. Plasma assisted milling of heat-resistant superalloys. Journal of Manufacturing Science and Engineering. **126**(2): 274–285. DOI: 10.1115/1.1644548

[127] Kim J-H, Lee C-M, Kim D-H. The effect of plasma-assisted machining and additive path strategies of Inconel 718 manufactured with directed energy deposition. Journal of Materials Research and Technology. 2022;**19**: 1658–1672. DOI: 10.1016/j.jmrt.2022.05.108

[128] Choi M-G, *et al.*, “Effect of plasma input power and feeding speed on the surface temperatures of stainless steel 316L during plasma assisted milling,” in *2020 IEEE International Conference on Plasma Science (ICOPS)*, Dec. 2020, pp. 613–613. DOI: 10.1109/ICOPS37625.2020.9717652

[129] Lee Y-H, Lee C-M. 2019. A study on optimal machining conditions and energy efficiency in plasma assisted machining of Ti-6Al-4V. Materials. **12**(16):2590. DOI: 10.3390/ma12162590

[130] Kim S-G, Lee C-M, Kim D-H. Plasma-assisted machining characteristics of wire arc additive manufactured stainless steel with different deposition directions. Journal of Materials Research and Technology. 2021;**15**:3016–3027. DOI: 10.1016/j.jmrt.2021.09.130

[131] WANG S, WANG W, LIU Z, YANG D. Comparative research of plasma-assisted milling and traditional milling in synthesizing AlN. Plasma Science and Technology. 2017; **19** (6):064005. DOI: 10.1088/2058-6272/aa62f8

Chapter 3

Microstructural Evolution Mechanism in Fabrication and Micromachining of Selective Laser Melted Al Alloys

Hansong Ji, Chaojiang Li, Yuxin Yang, Heng Liu, M. Muneeb Waheed and Na Zhang

Abstract

This chapter introduces the microstructural evolution behaviors in the fabrication and micromachining processes of selective laser melted Al alloys, including the crystallographic characteristics of SLMed Al alloys, the impact of SLM process parameters and heat treatments on crystallographic microstructure, the influence of microstructure on machinability, the microstructural evolution in the machining process, and the impact of micromachining parameters on the evolution behaviors. The phase composition, grain size, grain orientation, and grain boundary are discussed.

Keywords: microstructure, evolution mechanism, SLM, fabrication, micromachining

1. Introduction

Additive manufacturing (AM) is a promising technique in the fields of aerospace, automotive, and architectural design. AM enables the fabrication of highly complex components with high dimensional accuracy [1]. This process eliminates the need for molds, minimizes material waste, and accelerates the production process. Different AM methods have different forming mechanisms, and the processed materials exhibit different characteristics. In the broader category of metal AM processes, selective laser melting (SLM) is particularly valued for its ability to produce components with high density, excellent mechanical properties, and relatively uniform microstructure, making it widely adopted in advanced applications.

Among metal materials, Al–Si-based alloys have become an outstanding choice for SLM owing to their advantageous combination of physical and mechanical properties. These alloys exhibit excellent meltability, which facilitates reliable interlayer bonding during the rapid melting and solidification cycles in SLM processes [2]. Additionally, their good hardenability ensures that the as-built microstructure can achieve favorable mechanical strength without requiring extensive post-processing. The inherent corrosion resistance of Al–Si alloys further enhances their suitability for demanding environments, while their high thermal conductivity supports efficient heat dissipation during both processing and service life [3]. Within this family, the AlSi10Mg alloy has distinguished itself as particularly well suited for AM due to its superior printability and robust mechanical properties. The addition of magnesium contributes to precipitation strengthening, enhancing the alloy's strength without significantly compromising ductility [4]. Notably, AlSi10Mg exhibits a high strength-to-weight ratio, a critical attribute for applications where minimizing mass while maintaining structural integrity is paramount. These qualities have led to its extensive adoption in aerospace industries, where lightweight, high-performance components are essential for improving fuel efficiency, reducing emissions, and enabling innovative design solutions. Consequently, AlSi10Mg is frequently employed in the manufacture of structural brackets, engine components, and complex aerodynamic parts using SLM technology, underscoring its strategic importance in advancing AM for critical engineering applications. In addition, by adding nanoparticles such as TiC [5], SiC [6], and TiN [7], aluminum-based composites (AMCs) [8] are formed that integrate the excellent mechanical properties of the aluminum matrix and nanoparticle-reinforced materials, thereby improving the performance of parts [9]. The application of SLM-produced AlSi10Mg in the Airbus A350XWB program allowed the consolidation of over 30 individual components of the vertical tail bracket into a single structure and reduced the production time from 30 days to just 19 hours. Similarly, antenna brackets on the latest satellites achieved a 40% reduction in weight by leveraging SLM with AlSi10Mg.

The application of SLM to aluminum alloys, while promising, presents a range of significant challenges intrinsically linked to their unique physical and chemical properties, limiting its widespread employment in industries. Characteristics such as low bulk density, poor powder flowability, high propensity for oxidation, elevated laser reflectivity, and superior thermal conductivity collectively complicate the melting, solidification, and overall fabrication process, which inevitably results in defects, including spheroidal inclusions, internal porosity, and residual unpurified inclusions [10]. The presence of such defects detrimentally impacts critical performance by reducing mechanical strength, inducing stress concentrations, compromising dimensional accuracy and surface finish quality, accelerating crack initiation that undermines fatigue resistance, and deteriorating corrosion behavior. In the study of SLMed Al alloys and subsequent microscale machining processes, understanding the evolution of microstructural features plays a critical role in enhancing the performance of fabricated components. The rapid solidification and directional heat input inherent to the SLM process often result in non-equilibrium microstructures – such as columnar grains, elemental segregation at grain boundaries, and porosity – which significantly influence the material's deformation behavior, surface integrity, and residual stress during post-processing [11], particularly during precision cutting. By investigating the microstructural evolution across AM, heat treatment, and micromachining stages, it becomes possible to tailor grain size, crystallographic texture, and the distribution of strengthening phases. Such control enables significant improvements in mechanical properties, surface quality, and service reliability, thereby providing a fundamental basis for the reliable application of SLMed Al alloys in advanced manufacturing fields.

Traditional metallurgical theories, largely developed within the context of conventional casting and forging, fall short in fully explaining the complex microstructure formation mechanisms in SLMed Al alloys. This gap highlights the imperative need for novel predictive frameworks and mitigation strategies tailored specifically to the AM context. Central to this endeavor is the comprehensive characterization of the fine microstructural features that develop during and after SLM processing. Detailed microstructural analyses – including grain morphology, phase distributions, defect typologies, and interfacial characteristics – provide critical insights into the fundamental physical phenomena governing defect nucleation, growth, and interaction with the matrix. The distinct crystallographic microstructures and defect characteristics inherent to SLMed Al alloys are fundamental to the performance discrepancies observed between AMed and conventionally processed Al components. A thorough grasp of the above-mentioned microstructural features is crucial to the widespread adoption of SLMed Al alloys in industrial applications. Specifically, three key processes govern the microstructure and resultant performance of SLMed parts: the AM process itself, subsequent post-processing, and necessary subtractive finishing operations. The additive stage plays a pivotal role, as its parameters critically influence melt pool dynamics, solidification rates, and grain growth behaviors [3]. The post-thermal treatments, such as aging or solution treatment, are employed to promote precipitation strengthening, relieve residual stresses, homogenize the microstructure, and reduce supersaturated solid solutions and non-equilibrium phases [12]. Finally, due to the inherently high surface roughness and limited dimensional precision of SLMed components, subtractive processes – such as micromachining – are essential [13]. These finishing steps not only improve geometric accuracy and surface quality but also remove surface-connected defects such as pores, thereby enhancing fatigue resistance and overall service reliability.

Extensive research has been conducted on the microstructural evolution and process optimization of SLMed Al alloys. By integrating the results of these studies and clarifying the evolution mechanisms, reliable references can be provided for other researchers and industrial applications. This chapter is devoted to offering a detailed insight into the microstructural evolution behaviors in the fabrication and micromachining of SLMed Al alloys, including three aspects, that is, SLM, heat treatment, and micromachining processes.

2. Microstructural evolution in SLM of Al alloys

2.1 SLM parameter effects on the microstructure of SLMed Al alloys

The controlling parameters of the SLM process can be divided into four types, that is, laser-related, scan-related, powder-related, and temperature-related parameters, as listed in **Figure 1** [10].

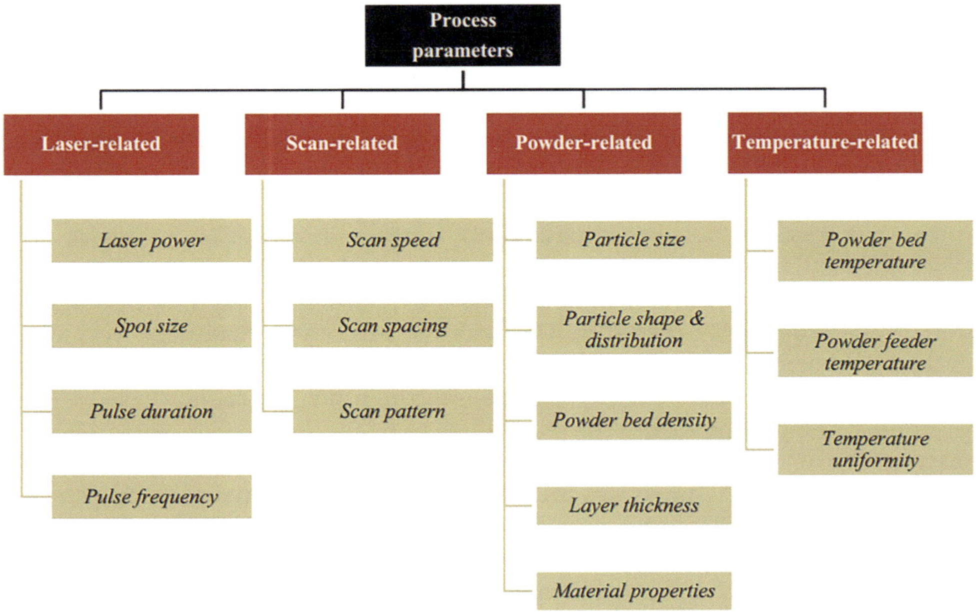

Figure 1.
Process parameters in the SLM process.

2.1.1 Laser energy density

During SLM processes, the most influential parameters are laser power (P), scanning speed (v), scanning spacing (h), and laser thickness (t), which demonstrate not only independent impacts but also coupling effects. In order to estimate the synthetic effects, a parameter named laser energy density, E, was created [14]:

$$E = \frac{P}{v \cdot h \cdot t} \tag{1}$$

The units of E, P, v, h, and t are, respectively, J/mm^3, J/s, mm/s, mm, and mm.

In the SLMed AlSi10Mg, the microstructure is primarily composed of α-Al phase and Al–Si eutectic phases. According to the structural features of eutectic phases, the microstructure of a single-track melted pool can be divided into three regions [15], that is, fine zone, coarse zone, and heat-affected zone (HAZ) [16, 17], as shown in **Figure 2**. In the fine zone, the microstructure of eutectic phases is mesh-shaped with a clear direction pointing from the pool boundary to the pool center. In the coarse zone, the microstructure is also mesh-shaped, while the orientation is not significant. In the HAZ, the microstructure of eutectic phases shows a scattered, granular shape. The results of Wu et al. [18] show that, with the increase in laser energy density from 0.27 to 0.53, the size of eutectic phases decreases, the aspect ratio increases, and the fraction of eutectic phases increases. According to the characteristics of grains, three melting modes can be extracted in this range, that is, the conduction mode, the mixed mode, and the

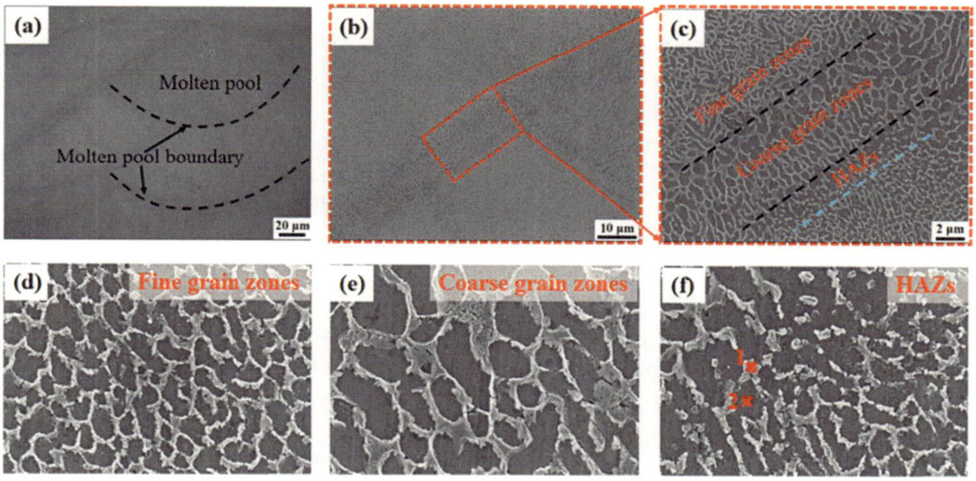

Figure 2.
Three different regions in the microstructure of SLMed AlSi10Mg [17].

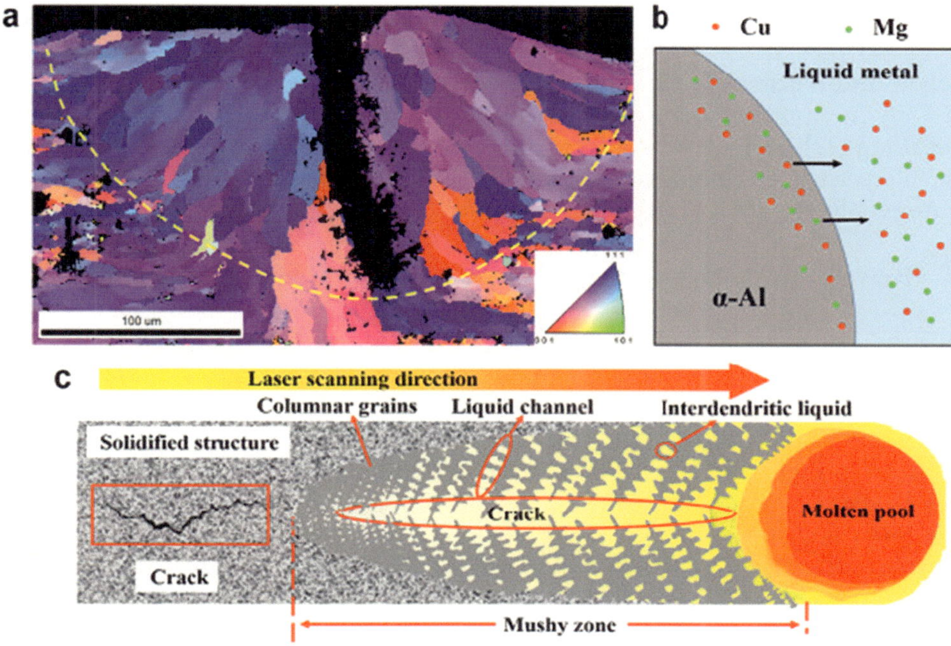

Figure 3.
The grains and defects in the keyhole mode [19].

keyhole mode. In the conduction mode, the grains appear columnar-shaped with their orientation perpendicular to the pool boundary and pointing to the pool center. With the increase in laser energy density, the conduction mode transitions to the mixed mode, and the shape of grains gradually transitions from columnar to equiaxed. As the laser energy density further increases, the mixed mode transitions to the keyhole mode. In this mode, the depth of the pool is deep, and vaporization is more pronounced. The grains enlarge, the aspect ratio decreases, the crystallographic texture weakens, and the keyhole defects become visible, as shown in **Figure 3** [19]. As for specific process parameters, the increase in laser power, the decrease in scanning speed, scanning spacing, and layer thickness will increase the laser energy density, and in turn influence the microstructure of SLMed AlSi10Mg following the above-mentioned path.

2.1.2 Other SLM parameters

The laser spot size influences the cooling rates in the solidification process and therefore influences the crystallographic grains [20]. The smaller spot size results in finer grains. The grains in the sample under a 120 μm spot size are basically coarsened columnar tree-like crystals, while the grains in the sample under a 60 μm spot size are mainly finer columnar tree-like crystals with a few equiaxed grains. Besides, the grains under a larger spot size show strong texture in the (001) orientation, while the grains under a smaller spot size show weak texture. The texture intensity of the grains under a larger spot size is 4.8 times that of those under a smaller spot size.

The powder size is also important in SLM processes, which generally ranges from 15 to 53 μm. There is more gas in the layer of finer powder, resulting in more gas pores and lower density. The coarser powder is close to the spot size, leading to more lack of fusion defects and lower density as well [19].

Layer rotation angle has a significant impact on the microstructure of SLMed AlSi10Mg [21]. The increase in layer rotation angle significantly reduces the pores in the microstructure compared to 0° and 45°. Similarly, Si in the α-Al/Si eutectic is more evenly distributed, resulting in improved mechanical properties.

2.2 Post-treatment effects on the microstructure of SLMed Al alloys

Heat treatment is required for SLMed AlSi10Mg alloy to improve its mechanical properties, and the commonly used treatment processes are solution and aging. A typical microstructure of SLMed AlSi10Mg is shown in **Figure 4**, and the microstructure after solution treatment and T6 (solution + aging) treatments is shown in **Figure 5** [22]. After solution treatments (a ~ c) and T6 treatments (d ~ f), dispersed Si-particles and a few needle-like Al–Si–Fe precipitates can be found in the matrix, which is the origin of strengthened mechanical properties. Besides, with the increase in solution temperature (from a to c), the size of Si-particles increases. The aging process after the solution process further increases the size of particles. The increase in Si-particle size results in the improvement of mechanical properties.

2.3 Reinforcement phase effects on the microstructure of SLMed AMCs

The addition of the reinforcement phase will significantly improve the microstructure of SLMed AMCs. For an example, Zhang et al. [23] investigated the impact of the SiC reinforcement phase on AlSi10Mg under variable laser energy densities. Results show that SiC and Al_4SiC_4 are the primary reinforcing phases at

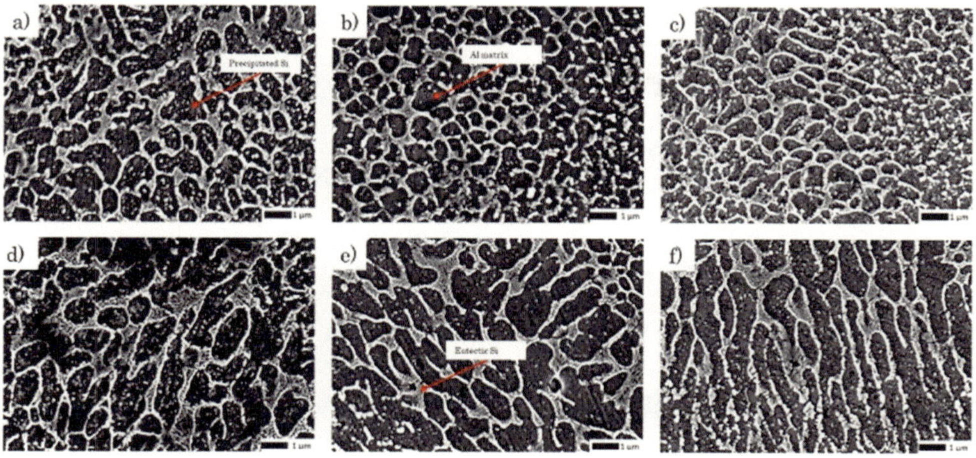

Figure 4.
A typical microstructure of SLMed AlSi10Mg [22].

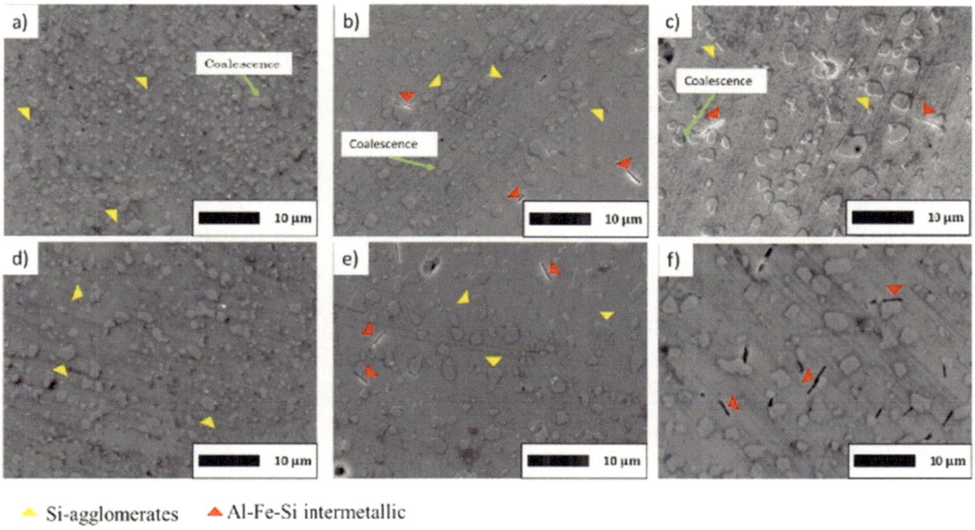

Figure 5.
The microstructure after heat treatments [22].

lower laser energy densities. With the increase in laser energy density, Al_4SiC_4 phase grows in a lamellar fashion. With further increases, Al_4SiC_4 phase grows sufficiently and forms a network distribution. Furthermore, the Al_4SiC_4 redistributes more uniformly in the Al matrix owing to a decrease in melt pool viscosity and an increase in surface tension gradient [24]. At higher laser densities, the unmelted SiC and Al_4SiC_4 particles disperse evenly in the microstructure due to Marangoni convection, which improves the microhardness [25].

3. The microstructural evolution of SLMed alloys in micromachining

3.1 Mechanisms of microstructural evolution in micromachining

Micromachining processes impose unique thermal and mechanical conditions that significantly influence microstructure evolution. Unlike conventional machining, micromachining operates at a scale where the key parameters become comparable to the grain size or even the overall deformation zone, thereby introducing dominant size effects [26].

3.1.1 Size effects

When the uncut chip thickness in mechanical micromachining becomes comparable to the grain size, the material removal process shifts from being primarily shear-dominated to being dominated by plowing-induced deformation [27]. This leads to higher specific cutting energy and poorer surface finish, particularly when chip formation becomes unstable or even impossible due to insufficient stress accumulation. At the microscale, high strain rates ($\geq 10^3\ s^{-1}$) result in dense dislocation networks and subgrain formation, facilitating grain refinement [28].

3.1.2 Thermal effects and dynamic recrystallization (DRX)

Thermal energy generated during microcutting can often exceed 300°C in the cutting zone, which can further drive microstructural changes. At elevated temperatures, DRX becomes significant, particularly in alloys such as AlSi10Mg, where DRX thresholds (~200–300°C) are relatively low [29]. This process results in the formation of fine, equiaxed grains that enhance mechanical uniformity [30].

3.1.3 Localized thermal softening and amorphization

Rapid thermal cycling from laser or μEDM may induce localized amorphization or ultrafine grains, although it can complicate phase stability. These microstructural transitions offer potential for tailored properties but demand process control.

In summary, microstructure evolution in micromachining is driven by a coupled effect of size-dependent deformation mechanisms, dislocation accumulation, and localized thermally induced recrystallization. Understanding these mechanisms is critical for tailoring the surface integrity and performance of machined Al alloys at the microscale.

3.2 Microstructural evolution in mechanical micromachining

Mechanical micromachining differs significantly from macroscale processes [31]. The minimal chip thickness, tool edge radius, and depth of cut all strongly influence the deformation and microstructural evolution behaviors in the material's near surface [32]. There is little research regarding the micromachining of SLMed Al alloys. Thus, this section considers other SLMed alloys.

3.2.1 Micromilling

Micromilling enables cost-effective fabrication of complex microcomponents [33], which is a promising technique to produce high-quality SLMed components.

Ji et al. [34] compared the influences of micromilling on SLMed and wrought Inconel 718 alloy. As shown in **Figure 6**, the unaffected zones represent the microstructure before micromilling, while the affected zones represent the

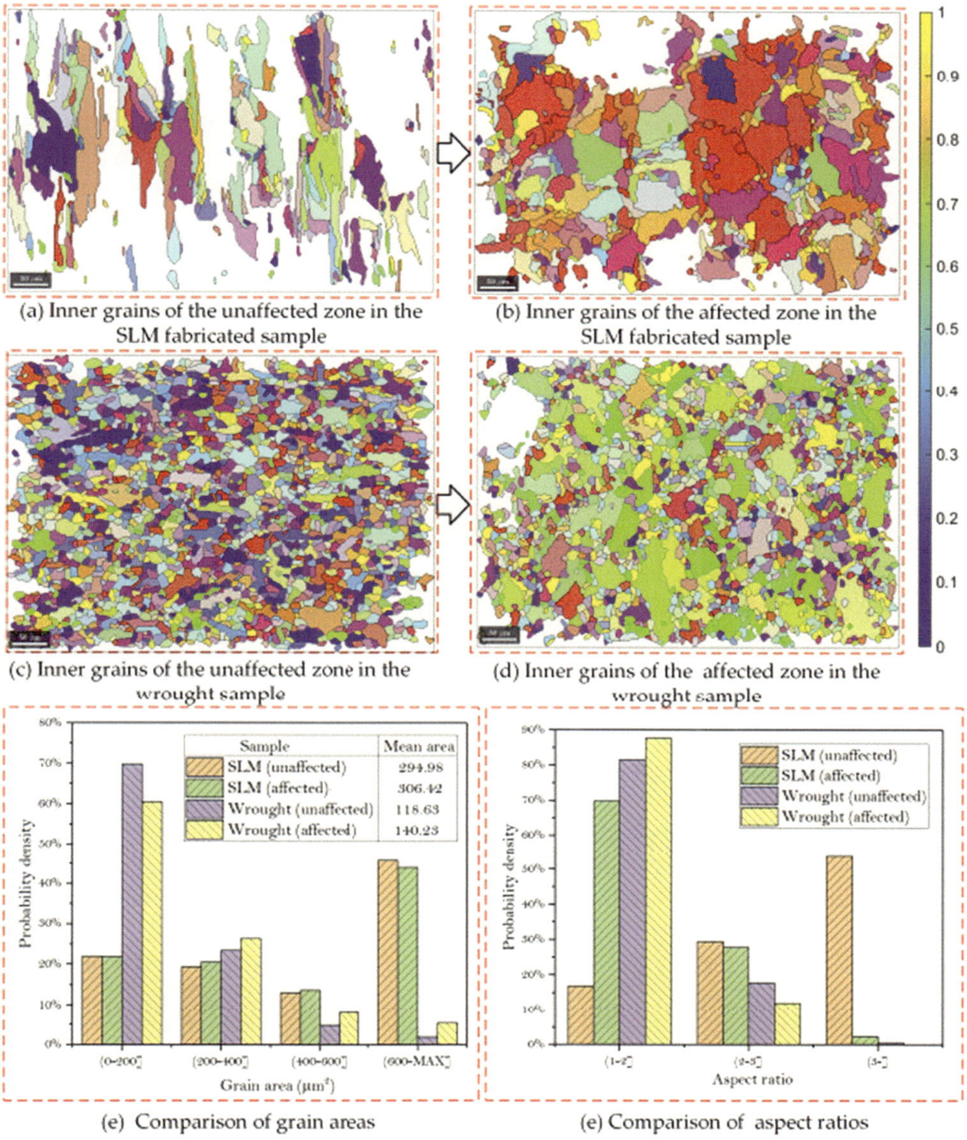

Figure 6.
Crystallographic microstructure comparison of SLMed and wrought Inconel-718 before and after micromilling.

microstructure after micromilling. The original microstructure of the wrought sample has fine equiaxed crystals, and that of the SLMed sample has large columnar dendritic grains. After micromilling, the grains in the affected area of the wrought sample exhibit a noticeable coarsening effect, with the average grain size increasing from 118.63 μm^2 to 140.23 μm^2. In contrast, the grains in the affected zone of the SLMed sample show only a slight coarsening, with the mean grain area rising from 294.98 μm^2 to 306.42 μm^2. This difference is primarily due to the fine, equiaxed crystal structure of the wrought sample, which leads to higher cutting forces and more heat generation in the cutting zone, unlike the larger columnar dendritic grains of the SLMed sample. As a result, thermally induced recrystallization plays a dominant role in the micromilling process of wrought IN718, while for SLMed IN718, severe plastic deformation-induced grain refinement and thermally induced recrystallization both contribute similarly.

Prior to micromachining, the SLMed sample exhibits a prominent <100> || building direction fiber texture with an intensity of 4.0. In contrast, the wrought sample displays four weaker fiber textures, each with an intensity of around 1.0. After micromilling, the original textures of the wrought sample diminished, and three new, weaker fiber textures emerged as a result of machining-induced changes. In comparison, the SLMed sample's initial texture weakened, and an additional <100> fiber texture with an intensity of 3.2 was generated.

As compared to the machinability of wrought AlSi10Mg, the SLMed AlSi10Mg exhibits lower compressive residual stress, less tool wear, better surface finish, and lower surface hardness.

3.2.2 Microgrinding

Microgrinding is commonly applied in the surface finishing of hard and brittle materials. Zhang et al. [35] explored the impact of microgrinding on the microstructure of SLMed FeCoNiCrAl0.5 high-entropy alloy. As the grinding speed increases, the interactions between the abrasive grains on the microabrasive tool and the SLMed FeCoNiCrAl0.5 workpiece intensify, leading to a thicker grinding metamorphic layer. Additionally, with a greater grinding depth, the thickness of the metamorphic layer also increases.

3.3 Effects of non-traditional micromachining methods

Non-conventional micromachining techniques, such as laser-based processing and ultrasonic vibration-assisted machining (UVAM), introduce unique thermal and mechanical environments, as lised in **Table 1**. These methods often involve high energy density, rapid thermal cycling, or stress wave propagation, which profoundly alter the microstructure of aluminum alloys beyond what conventional cutting can achieve.

3.3.1 Laser micromachining

Laser micromachining provides precise control at the micrometer scale over surface and shape, making it a promising post-processing method for SLMed components. Ghosh et al. [36] studied the effect of femtosecond laser

Process	Parameter	Effect on machining outcomes
Laser micromachining	Laser power [39]	Higher power causes a wider kerf, a larger HAZ, and increased surface roughness.
	Scanning speed [39]	Higher speed causes reduced kerf width, HAZ, and surface roughness.
	Pulse frequency [39]	Non-linear effect on roughness: Initial increase, then decrease.
Laser surface treatment	Spot spacing [40]	Smaller spacing (< spot diameter) causes micro-nano composite structures and enhanced adhesion.
Laser shock peening	Intensity [41–43]	Higher intensity causes grain refinement, compressive residual stresses, and a 15–30% reduction in porosity.
UVAM	Vibration frequency [2, 37]	1) 20 kHz causes reduced friction and refined equiaxed grains. 2) 969 Hz causes 99.1% density and grain size refinement (from 4–13 μm to 2–4 μm).
	Vibration application [37]	UVAM causes surface roughness (11.03 nm $\rightarrow$ 5.10 nm) and cutting forces (21.86–27.62%).

Table 1.
Non-conventional micromachining parameters and effects.

micromachining on the microstructure of SLMed SS316L microstruts. The surface of the as-built microstrut consists of equiaxed grains, while the core features elongated columnar grains. The inverse pole figure map shows a transition in texture, shifting from a preferred <110> crystallographic direction in the core to a random texture at the surface. As the grains move closer to the surface, the <110> texture deviates until it becomes entirely random. In contrast, laser micromachining removes the surface grains from the as-built state, leaving only the <110> core grains. Consequently, the micromachined specimens display a dominant <110> texture throughout their microstructure.

3.3.2 Ultrasonic vibration-assisted machining (UVAM)

UVAM applies high-frequency (~20 kHz) vibration to enhance chip breaking, reduce friction, and refine microstructure. It promotes dynamic recovery, subgrain formation, and grain refinement, with EBSD revealing equiaxed grains and weaker texture for improved isotropy [37]. Surface roughness decreases from 11.03 nm to 5.10 nm, while cutting forces are reduced by up to 27.6% [37]. High-frequency microvibration (969 Hz) in laser AM further refines TiC/AlSi10Mg alloys, reducing pores (99.1% density) and grain size (2 ~ 4 μm vs. 4 ~ 13 μm), achieving 183.3 HV0.05 hardness and 314.7 MPa tensile strength. **Figure** 7 [38] illustrates that ultrasonic vibration during WAAM promotes the transformation from coarse columnar to fine equiaxed grains, significantly reducing grain size. Excessive amplitude, however, may cause grain coarsening and cracking.

Non-conventional microfabrication techniques utilize energy fields (e.g., laser, ultrasound) to precisely control material transformations. As shown in the table, parameters such as energy density and vibration frequency critically affect defect reduction, grain refinement, and surface properties, enabling microstructure-by-design approaches.

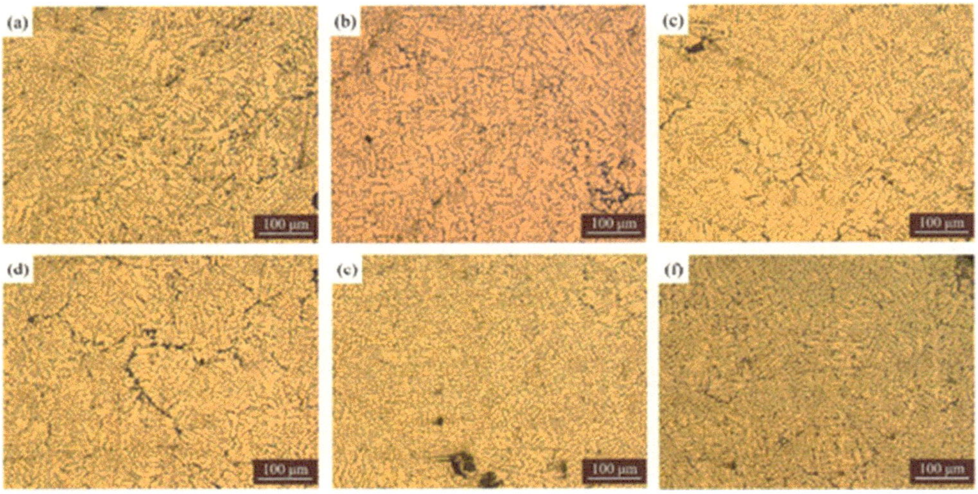

Figure 7.
Microstructure of WAAM samples with and without ultrasonic vibration [38].

Acknowledgments

This work was supported by the National Natural Science Foundation of China (Grant Numbers: 52305481 and 52375402).

Author details

Hansong Ji, Chaojiang Li*, Yuxin Yang, Heng Liu, M. Muneeb Waheed and Na Zhang
Beijing Institute of Technology, Beijing, China

*Address all correspondence to: mecjli@bit.edu.cn

References

[1] Jiang P, Ji Z, Yan C, Wang X, Zhou F. High compressive strength metallic architectures prepared via polyelectrolyte-brush assisted metal deposition on 3D printed lattices. Nano-Structures and Nano-Objects. 2018;**16**:420–427.

[2] Li C, Sun S, Liu C, Lu Q, Ma P, Wang Y. Microstructure and mechanical properties of TiC/AlSi10Mg alloy fabricated by laser additive manufacturing under high-frequency micro-vibration. Journal of Alloys and Compounds. 2019;**794**:236–246.

[3] Gao C, Wu W, Shi J, Xiao Z, Akbarzadeh AH. Simultaneous enhancement of strength, ductility, and hardness of TiN/AlSi10Mg nanocomposites via selective laser melting. Additive Manufacturing. 2020;**34**:101378.

[4] Kang N, Coddet P, Dembinski L, Liao H, Coddet C. Microstructure and strength analysis of eutectic Al-Si alloy *in-situ* manufactured using selective laser melting from elemental powder mixture. Journal of Alloys and Compounds. 2017;**691**:316–322.

[5] Jia Q, Gu D. Selective laser melting additive manufacturing of TiC/inconel 718 bulk-form nanocomposites: Densification, microstructure, and performance. Journal of Materials Research. 2014;**29**(17):1960–1969.

[6] Song J, Guo Q, Ouyang Q, Su Y, Zhang J, Lavernia EJ, et al. Influence of interfaces on the mechanical behavior of SiC particulate-reinforced Al–Zn–Mg–Cu composites. Materials Science and Engineering: A. 2015;**644**:79–84.

[7] Zhao X, Wei QS, Gao N, Zheng EL, Shi YS, Yang SF. Rapid fabrication of TiN/AISI 420 stainless steel composite by selective laser melting additive manufacturing. Journal of Materials Processing Technology. 2019;**270**:8–19.

[8] Martin JH, Yahata BD, Hundley JM, Mayer JA, Schaedler TA, Pollock TM.3D printing of high-strength aluminium alloys. Nature. 2017;**549**(7672):365–369.

[9] Orme ME, Gschweit M, Ferrari M, Vernon R, Madera IJ, Yancey R, et al. Additive manufacturing of lightweight, optimized, metallic components suitable for space flight. Journal of Spacecraft and Rockets 2017;**54**:1050–1059.

[10] Aboulkhair NT, Everitt NM, Ashcroft I, Tuck C. Reducing porosity in AlSi10Mg parts processed by selective laser melting. Additive Manufacturing. 2014;**1–4**:77–86.

[11] Rong T, Gu D. Formation of novel graded interface and its function on mechanical properties of WC1-x reinforced inconel 718 composites processed by selective laser melting. Journal of Alloys and Compounds. 2016;**680**:333–342.

[12] B S, B S. A comparison of fatigue strength sensitivity to defects for materials manufactured by AM or traditional processes. International Journal of Fatigue. 2017;**94**:178–191.

[13] Cevik ZA, Ozsoy K, Ercetin A, Sariisik G. Machine Learning-Driven Optimization of Machining Parameters Optimization for Cutting Forces and Surface Roughness in Micro-Milling of AlSi10Mg Produced by Powder Bed

Fusion Additive Manufacturing. Applied Sciences. 2025;**15**(12):6553.

[14] Thijs L, Verhaeghe F, Craeghs T, Humbeeck JV, Kruth JP. A study of the microstructural evolution during selective laser melting of Ti–6Al–4V. Acta Materialia. 2010;**58**(9):3303–3312.

[15] Zhao L, Song L, Santos Macías JG, Zhu Y, Huang M, Simar A, et al. Review on the correlation between microstructure and mechanical performance for laser powder bed fusion AlSi10Mg. Additive Manufacturing. 2022;**56**:102914.

[16] Thijs L, Kempen K, Kruth JP, Van Humbeeck J. Fine-structured aluminium products with controllable texture by selective laser melting of pre-alloyed AlSi10Mg powder. Acta Materialia. 2013;**61**(5):1809–1819.

[17] Dong Z, Xu M, Guo H, Fei X, Liu Y, Gong B, et al. Microstructural evolution and characterization of AlSi10Mg alloy manufactured by selective laser melting. Journal of Materials Research and Technology. 2022;**17**:2343–2354.

[18] Wu A, Shi Y, Jing Y, Li S, Li W, Su X. Effects of linear energy density on melt pool and microstructure of selective laser melted AlSi10Mg: An experimental and numerical study. Materials Today Communications. 2025;**43**:111703.

[19] Yang Yongqiang WM, Trofimov V, Changhui S, Hanxiang Z, Wand D. Effects of particle size on processability of AlSi10Mg alloy manufactured by selective laser melting. Acta Metallurgica Sinica. 2023;**59**(1):147–156.

[20] Hung CH, Chu CW, Chang TL, Lin TC, Chen YP. Effect of focal spot size on AlSi10Mg alloy parts fabricated by laser powder bed fusion: Process window, mechanical properties, and microstructure. Journal of Alloys and Compounds. 2024;**977**:173338.

[21] Gupta MK, Singla AK, Ji H, Song Q, Liu Z, Cai W, et al. Impact of layer rotation on micro-structure, grain size, surface integrity and mechanical behaviour of SLM Al-Si-10Mg alloy. Journal of Materials Research and Technology 2020;**9**:9506–9522.

[22] Syrlybayev D, Perveen A, Talamona D. Controlling mechanical properties and energy absorption in AlSi10Mg lattice structures through solution and aging heat treatments. Materials Science and Engineering: A. 2024;**889**:145843.

[23] Zhang D, Yi D, Wu X, Liu Z, Wang W, Poprawe R, et al. SiC reinforced AlSi10Mg composites fabricated by selective laser melting. Journal of Alloys and Compounds. 2022;**894**:162365.

[24] Gao C, Wang Z, Xiao Z, You D, Wong K, Akbarzadeh AH. Selective laser melting of TiN nanoparticle-reinforced AlSi10Mg composite: Microstructural, interfacial, and mechanical properties. Journal of Materials Processing Technology. 2020;**281**:116618.

[25] Chang F, Gu D, Dai D, Yuan P. Selective laser melting of in-situ Al4SiC4 + SiC hybrid reinforced Al matrix composites: Influence of starting SiC particle size. Surface and Coatings Technology. 2015;**272**:15–24.

[26] Balázs BZ, Takács M. Experimental investigation and optimisation of the micro milling process of hardened hot-work tool steel. The International

Journal of Advanced Manufacturing Technology. 2020;**106**(11-12):5289–5305.

[27] Dib MHM, Duduch JG, Jasinevicius RG. Minimum chip thickness determination by means of cutting force signal in micro endmilling. Precision Engineering. 2018;**51**:244–262.

[28] Al-Zubaydi ASJ, Gao N, Wang S, Reed PAS. Microstructural and hardness evolution of additively manufactured Al–Si–Cu alloy processed by high-pressure torsion. Journal of Materials Science. 2022;**57**(19):8956–8977.

[29] Zhang D, Zhang XM, Leopold J, and Ding H Subsurface deformation generated by orthogonal cutting: Analytical modeling and experimental verification. Journal of Manufacturing Science and Engineering [Internet]. [cited 2025 Jul 13];**139**(9). Available from: https://asmedigitalcollection.asme.org/manufacturingscience/article/doi/10.1115/1.4036994/477256/Subsurface-Deformation-Generated-by-Orthogonal [Accessed: 2017]

[30] Guang CC, hao HW, Qi M, Peng DS, Li P, Yang F, et al. Microstructural evolution and mechanical properties of an ultrahigh-strength Al-Zn-Mg-Cu alloy via powder metallurgy and hot extrusion. Journal of Central South University 2021;**28**:1195–1205.

[31] Liu X, DeVor RE, Kapoor SG, Ehmann KF. The Mechanics of Machining at the Microscale: Assessment of the Current State of the Science. Journal of Manufacturing Science and Engineering. 2004;**126**(4):666–678.

[32] Comprehensive Materials Processing. In: Comprehensive Materials Processing [Internet]. Elsevier; [cited 2025 Jul 13]. p. iii. Available from: https://linkinghub.elsevier.com/retrieve/pii/B9780323960205099027 [Accessed: 2024]

[33] Sun Y, Liang Y, and Cheng K. Micro-Scale And Meso-Scale Mechanical Manufacturing. China Mechanical Engineering. 2004;**40**(5):1–6.

[34] Ji H, Gupta MK, Song Q, Cai W, Zheng T, Zhao Y, et al. Microstructure and machinability evaluation in micro milling of selective laser melted inconel 718 alloy. Journal of Materials Research and Technology. 2021;**14**:348–362.

[35] Zhang WB, Wen XL, Song LY, Gong YD, Xu RC. Experimental study on surface quality in micro-grinding FeCoNiCr high-entropy alloy prepared by selective laser melting. Archives of Civil and Mechanical Engineering. 2023;**24**(1):27. DOI: 10.1007/s43452-023-00834-0.

[36] Ghosh A, Biswas S, Turner T, Kietzig AM, Brochu M. Surface, microstructure, and tensile deformation characterization of LPBF SS316L microstruts micromachined with femtosecond laser. Materials & Design. 2021;**210**:110045.

[37] Bai Y, Shi Z, Lee YJ, Wang H. Optical surface generation on additively manufactured AlSiMg0.75 alloys with ultrasonic vibration-assisted machining. Journal of Materials Processing Technology. 2020;**280**:116597.

[38] Zhang J, Xing Y, Zhang J, Cao J, Yang F, Zhang X. Effects of In-Process Ultrasonic Vibration on Weld Formation and Grain Size of Wire and Arc Additive Manufactured Parts. Materials. 2022;**15**(15):5168

[39] Mandal KK, Kuar AS, Mitra S. Experimental investigation on laser micro-machining of Al 7075 alloy. Optics and Laser Technology. 2018;**107**:260–267.

[40] Li G, Lei M, Liang C, Li K, Li B, Wang X, et al. Improving adhesive bonding of Al alloy by laser-induced micro–nano structures. Applied Sciences 2022;**12**:1199

[41] Salimianrizi A, Foroozmehr E, Badrossamay M, Farrokhpour H. effect of laser shock peening on surface properties and residual stress of Al6061-T6. Optics and Lasers in Engineering. 2016;**77**:112–117.

[42] Damon J, Dietrich S, Vollert F, Gibmeier J, Schulze V. Process dependent porosity and the influence of shot peening on porosity morphology regarding selective laser melted AlSi10Mg parts. Additive Manufacturing. 2018;**20**:77–89.

[43] Maamoun AH, Elbestawi MA, Veldhuis SC. Influence of shot peening on AlSi10Mg parts fabricated by additive manufacturing. Journal of Manufacturing and Material Processing. 2018;**2**(3):40.

Section 3

Surface Modification and Surface Engineering

Chapter 4

Plasma Nitriding of Al–Cu and Al–Cu–Mg Alloys at Various Holding Temperature

Tatsuhiko Aizawa, Hiroki Nakata and Takeshi Nasu

Abstract

A plasma nitriding has been highlighted as a surface treatment to harden aluminum alloys and improve their wear resistance. Various processing approaches were reviewed to describe the nitriding processes at relatively high temperatures. Duralumin and super-duralumin alloys, such as Al–Cu and Al–Cu–Mg alloys, were nitrided by DC- and RF-DC plasmas at holding temperatures ranging from 723 K to 623 K. An AlN (Aluminum Nitride)-rich layer formation characterized the nitriding behavior of Al–Cu alloys at 673 K. The intermetallic precipitate of Al_2Cu facilitated this fast-rate AlN-layer formation as a catalyst. This process was affected at 723 K by Mg diffusion and surface oxidation in the case of plasma nitriding of Al–Cu–Mg alloys. The plasma nitriding at 623 K was free from this deterioration, allowing the AlN-rich layer to form in a manner similar to the nitriding of Al–Cu alloys. This layer exhibited a hardness greater than 1,000 HV, which is 10 times higher than the original hardness of Al–Cu–Mg alloys. A heatsink was plasma nitrided at 623 K to demonstrate that aluminum alloy parts could be homogeneously nitrided to form the AlN-rich layer. The plasma nitriding at temperatures lower than 600 K was investigated to explore the possibility of nitrogen supersaturation in aluminum alloys, as well as the accommodated nitriding process via intermetallic precipitates other than Al_2Cu.

Keywords: plasma nitriding, intermetallic precipitates, AlN-rich layer formation, surface hardness, holding temperature effect, duralumin alloys

1. Introduction

Light-weight metals and alloys have been highlighted as structural materials for electric vehicles, rail-road trains, and tracks to save energy consumption and reduce environmental burdens [1]. Among them, aluminum and aluminum alloys are worth using as sustainable materials due to their recyclability with minimum energy consumption [2]. In the life cycle assessment of aluminum and aluminum alloy products, their reusability must be improved by increasing wear resistance via surface treatment [3]. Anodizing process [4] was invented to synthesize an alumina layer onto the surface of various aluminum products. In addition to its residual porosities, its low toughness has been identified as a fatal drawback that shortens the life time of aluminum alloy products.

Instead of the oxidation process, various nitriding treatments have been proposed to harden steels [5]; e.g., as listed in **Table 1**, the gas nitriding using ammonia [6], liquid nitriding using chloride solutions [7], and plasma nitriding using a nitrogen–hydrogen mixture gas [8]. Among these, plasma nitriding provides an environmental-friendly method to improve surface hardness and tribological performance, as compared in **Table 1**.

In the literature, various methods have also been proposed for the surface treatment of aluminum and aluminum alloys. **Table 2** compares three plasma nitriding processes used to form a nitrided surface layer on aluminum and aluminum alloys. Relatively high holding temperatures and longer nitriding durations were

Nitriding	Source	TemperatureK	Durationks	HardnessHV	Thicknessμm	Work
Salt bath nitriding	NaCN KCN	833–853	180–360	1000	100–300	Alloy steel
Gas nitriding	NH_3	733–853	90–360	700–1,200	100–800	High alloy steels
Plasma nitriding	$N_2 + H_2$	623–863	3.6–108	600–1,300	50–90	Steels

Table 1.
Comparison of three nitriding processes for surface treatment of steels.

Method	Performance	Merits	Demerits	REF
Ion implantation	Implantation energy (>10KeV) Lattice distension of AlN	No need for presputtering	High energy consumption Difficulty in application to complex-shaped parts Surface damage	[9, 10]
Plasma immersion Ion Implantation	10–15 μm thick film 723–823 K More than 18 ks 4.5 μm by 43.2 ks	Short processing time Thick AlN layer	Presputtering (36ks) Surface damage High energy consumption Difficulty in application to automotive parts	[11, 12]
Plasma nitriding	Several microns in thickness More than 723 K Long than 28.8 ks	Low energy consumption Low gas consumption Direct application to industry	Pre-sputtering by Ar Long nitriding time Relatively high holding temperature	[13, 14]

Table 2.
Comparison of three plasma nitriding processes for aluminum and aluminum alloys.

Processing	System	Features
DC-plasma processing	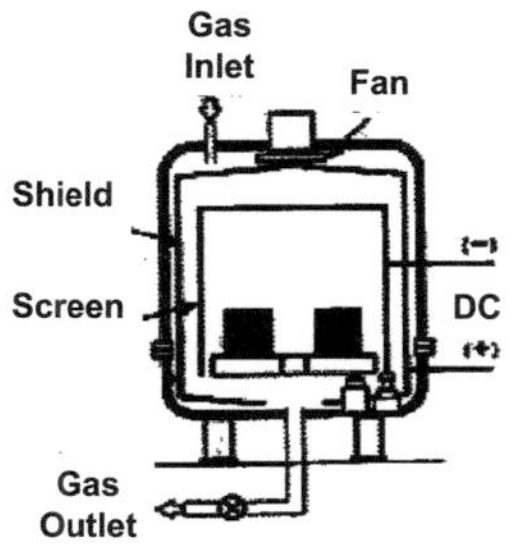	A few trials were reported on the nitriding of aluminum alloys by screen plasma technique.
RF-plasma processing	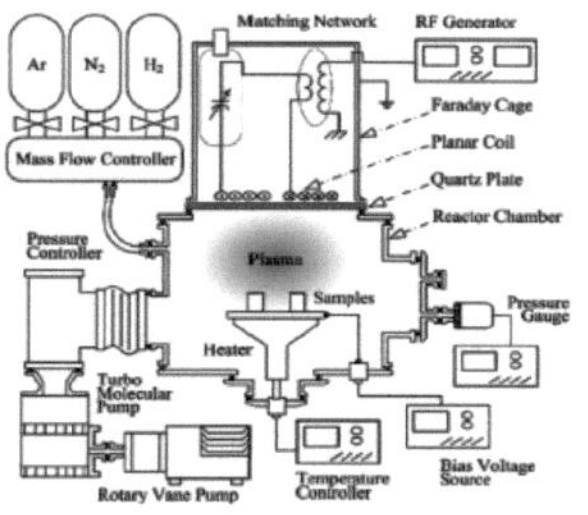	Induction-couples plasma (13.56 MHz) was utilized to nitride Al-6Cu alloys. Thin AlN layer was synthesized and detected by XPS
Electron beam assisted plasma processing	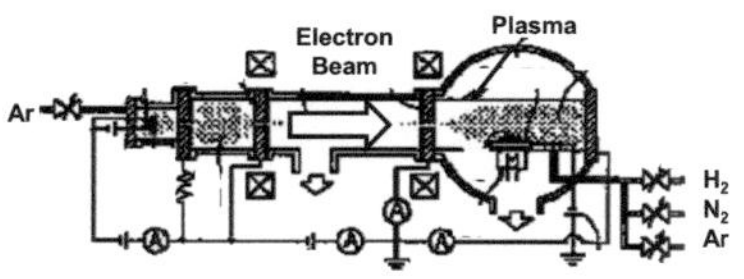	Plasma was enhanced by high current electron beams in low pressure: e.g., I > 10A, ~100 eV, 0.1 Pa

Table 3.
Comparison of three plasma nitriding processes for the advancement of nitriding characteristics.

common to these three processes. This fatal drawback might stem from the very low diffusion coefficient of nitrogen solutes through AlN [8].

In the literature, many efforts have also been made to modify the normal nitriding process for the synthesis of a thick AlN layer at a lower holding temperature and for a shorter duration. Three processes have been cited in the literature to describe this improvement in nitriding characteristics. As listed in **Table 3**, DC (direct current)-plasma nitriding, with the use of screens, played a role in nitriding various steels; however, little data has been reported on further improvement over the normal plasma nitriding described in **Table 2** [15]. RF (Radio Frequency)-plasma nitriding succeeded in lowering the holding temperature, and a very thin AlN-layer was synthesized using this approach [16]. With the assistance of the electron beam, plasma nitriding also succeeded in the synthesis of a thick AlN-layer at a relatively lower holding temperature and for a shorter duration [17].

In the present chapter, the nitriding behavior is reconsidered to search for an efficient plasma nitriding treatment at a lower holding temperature for a shorter nitriding duration. DC-plasma and RF–DC plasma nitriding processes are utilized to describe the nitriding behavior of Al–Cu and Al–Cu–Mg alloys in the 2000-series of aluminum alloys. The role of intermetallic precipitate (IMP) is discussed as the main mechanism driving the formation of a thick AlN-layer in both the DC- and RF–DC plasma nitriding processes. Microstructure analyses are performed to experimentally prove this Al_2Cu-accommodated formation of the AlN layer via the reactions by $Al_2Cu + 2N$ (diffusing nitrogen solute atoms) → 2AlN + Cu, and, Cu + 2Al (in matrix) → Al_2Cu. Al_2Cu works as a catalyst to drive the overall reaction by Al (in matrix) + N (diffusing nitrogen solute atom) → AlN. This reaction takes place at the interfacial boundaries between AlN and Al_2Cu synthesizing a thick composite layer of AlN and Al_2Cu within the Al–Cu and Al–Cu–Mg alloy matrices.

In addition to this positive role of IMP, the alloying element in these alloys influences the nitriding behavior. In the case where the Al–Cu–Mg alloys are RF–DC plasma nitrided at 723 K, the surface diffusion of Mg and its oxidation hinder the nitriding process. This deterioration is suppressed by lowering the temperature to 623 K to make AlN-layer formation active in nitriding the Al–Cu–Mg alloys. A duralumin heat sink is successfully nitrided at 623 K, demonstrating that aluminum alloy parts are hardened by the formation of an AlN-layer.

Two RF–DC plasma nitriding approaches are discussed for 5,000, 6,000, and 700 series aluminum alloys, other than Al–Cu and Al–Cu-Mg alloys, at low holding temperatures. A massive nitrogen supersaturation provides a candidate treatment to build up the AlN-layer using the in-situ granular refinement and fine grain- and cluster-boundary network for nitrogen diffusion. IMP, other than Al_2Cu works as a catalyst to drive the reaction by Al + N → AlN with less dependency on the chemical compositions of aluminum alloys.

2. Methods and materials

Two plasma nitriding systems were introduced to form AlN-rich surface layer on Al–Cu and Al–Cu-Mg alloys; e.g., DC plasma and RF–DC plasma nitriding systems. The standardized aluminum alloys, such as AA2011 or AA2014 were utilized as specimens, as well as the home-made alloys.

2.1 DC-plasma nitriding system

A DC-plasma nitriding system without the use of screens [15] was employed to form an AlN-rich layer on the Al–Cu and Al–Cu-Mg alloy specimens, which were prepared by their own melting and casting processes. Their chemical components are listed in **Table 4**. As shown in **Figure 1a**, the nitrogen – hydrogen plasma was ignited by glow discharge under the applied DC bias voltage. The specimen was set up on the DC-biased plate, below which the electrical heater was located. A thermocouple was used to measure the specimen temperature, using a dummy specimen [18].

A typical nitriding condition is listed in **Table 5**. To be discussed later, nitrogen and hydrogen gas pressures are fixed to be high enough to sustain the

Cu	Mg	Al	
Al-Cu alloy	6.0	0.0	Bal.
Al-Cu-Mg alloy	6.0	0.5	Bal.

Table 4.
Chemical compositions of Al–Cu and Al–Cu–Mg alloy specimens.

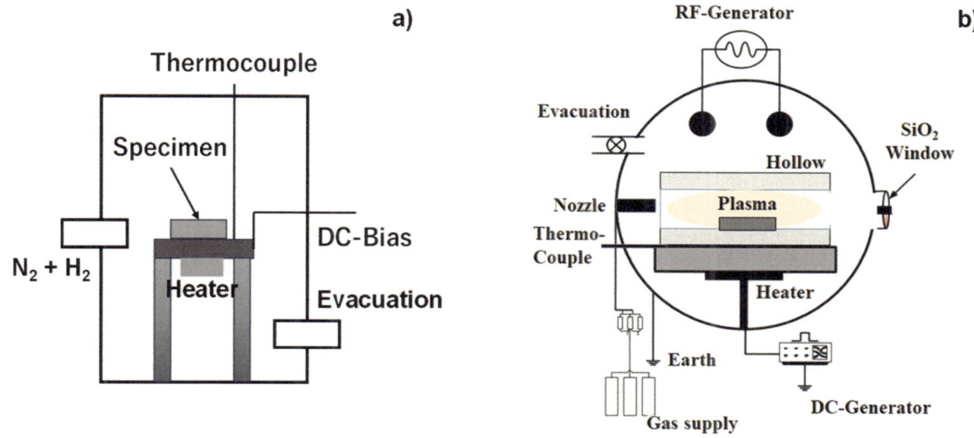

Figure 1.
Schematic views of two plasma nitriding systems. a) DC-plasma nitriding system, and b) RF–DC plasma nitriding system.

Process	Parameters
Pre-sputtering	DC (-500 V)Pressure (133 Pa)Temperature (723K)Duration (1.8 ks), Carrier gas (N_2 only)
Nitriding	DC (-500 V)Pressure (533 Pa)Temperature (673K)Duration (7.2, 14.4, 21.6, 28.8 ks)Carrier gas ($N_2 + H_2$) Partial pressure ratio (N_2:H_2 = 1 : 3)

Table 5.
DC-plasma nitriding conditions.

yield of NH-radicals in DC-plasma nitriding [19, 20]. In reference to ammonia gas, which is usually employed in the so-called ion- and radical-nitriding processes in industries [21, 22], the gas flow rate in the mixture gas was also fixed to be 25% for nitrogen gas and 75% for hydrogen gas.

2.2 RF–DC plasma nitriding system

A RF–DC plasma nitriding system using the hollow cathode was employed to describe the effect of holding temperature on nitriding behavior, as shown in **Figure 1b** [23]. A nitrogen–hydrogen plasma was ignited only in the hollow so that the nitrogen ion and NH-radical densities were enhanced to increase the

	Si	Fe	Cu	Mn	Mg	Cr	Zn	Ti	Ti+Zr	AI
A2011	0.15	0.23	5.42	0.00	0.00	0.00	0.01	0.03	———	Bal.
A2014	0.8	0.3	4.5	0.72	0.48	0.02	0.08	0.02	0.02	Bal.

Table 6.
Chemical compositions of AA2011 and AA2015 aluminum alloy specimens.

nitrogen solute flux into AA2011 or AA2014 alloy matrices. Their chemical compositions are listed in **Table 6**. A typical plasma nitriding condition is summarized in **Table 7**. Compared to the DC-plasma nitriding conditions in **Table 5**, lower gas pressure was utilized, around 75 Pa. The gas flow rate was determined by the in-situ plasma diagnosis to be 80% for nitrogen gas and 20% for hydrogen gas in mixture [24].

2.3 Mechanical characterization

Micro-hardness testing was employed to analyze the mechanical response of DC- and RF–DC plasma nitrided aluminum alloy specimens. The hardness depth profile was measured to describe the advancement of the nitrided layer into the depth of the matrix. The surface hardness, by varying the applied load, was also measured to analyze the hardening behavior due to AlN-formation during nitriding.

2.4 Microstructure characterization

SEM (scanning electron microscopy) – EDS (energy dispersive X-ray spectroscopy), as well as TEM (transmission electron microscopy), were utilized for microstructure characterization.

XRD (X-ray diffraction) and GIXD (glancing incident X-ray diffraction) were utilized to identify the synthesized phases during nitriding. XPS (X-ray photoelectron spectroscopy) was also used to analyze the synthesized compounds on the nitrided specimen surface, to describe the depth profile of related elements in the nitriding process, and to measure the binding state of compounds at the in situ-formed surface layer of the nitrided specimen.

Process	Parameters
Pre-sputtering	DC(-500 V), RF (0 V) Pressure (30 Pa) Temperature (623K, 673K, 723 K), Duration (1.8 ks), Carrier gas (N_2 only)
Nitriding	DC (-500 V), RF (250 V) Pressure (75 Pa) Temperature (623K, 673K, 723 K), Carrier gas (N_2 + H_2) Partial pressure ratio (N_2:H_2 = 4:1)

Table 7.
RF–DC plasma nitriding conditions.

3. DC-plasma nitriding of Al–Cu alloys at 673 K

DC-plasma nitriding was first employed to describe the nitriding behavior of Al–Cu alloys at 673 K. The nitriding duration was varied to examine the nitriding transients. An AlN-layer was formed as a hardened surface layer by DC-plasma nitriding for 28.8 ks. The nucleation and growth mechanism of the AlN-layer was experimentally analyzed to discuss the role of Al_2Cu precipitates in the AlN-layer formation.

3.1 Materials characterization

GIXD (grazing incident X-ray diffraction) was utilized to describe the nitriding process of Al–Cu alloys with increasing nitriding duration. As shown in **Figure 2**, the XRD profile changed significantly with duration. Peaks for aluminum nitride were detected with small intensity in the nitrided sample at 7.2 ks; the main peaks come from aluminum. This implies that AlN only nucleates along the grain boundaries. With increasing nitriding time, the peak intensity ratio of AlN to Al monotonically increases. At 14.4 ks, AlN and Al_2Cu grew to become the main components together with the aluminum matrix. After 14.4 ks, these aluminum peaks became relatively small, indicating that the nitrided layer consists of AlN and Al_2Cu.

Figure 3a depicts the surface of a DC-plasma nitrided Al–Cu alloy specimen at 673 K for 28.8 ks. The original aluminum alloy surface consists of a nitrided layer without the Al–Cu alloy matrix. This corresponds to the GIXD profile in **Figure 2,** where only AlN and Al_2Cu peaks are detected, without aluminum peaks. As shown in **Figure 3b**, the cross-section of the nitrided layer has a columnar structure. This thick nitrided layer has a columnar structure, and the length of this columnar structure is approximately equal to its thickness. The diameter of each column is around 20–30 μm, equivalent to the original grain size of the Al–Cu alloy. Through GIXD and EDX analyses, this column is composed only of AlN and Al_2Cu. This suggests that AlN nucleates along the grain boundaries and grows into the Al–Cu alloy matrix via the formation of a columnar composite of AlN and Al_2Cu.

The large yield of AlN in **Figures 2** and **3** is a proof that nitrogen atoms are supplied to the nitriding front end with sufficient nitrogen content to sustain the

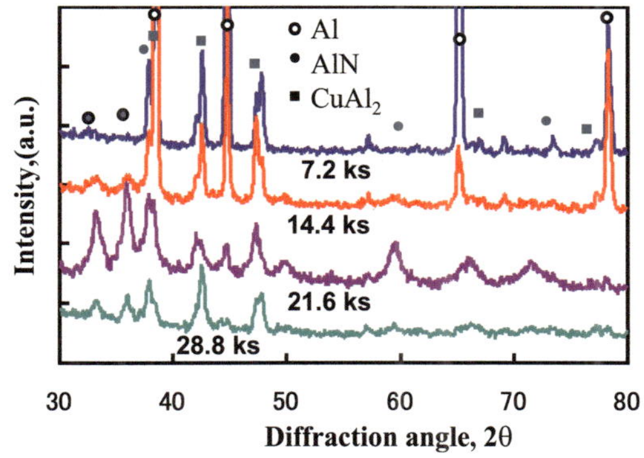

Figure 2.
Variation of GIXD diagrams for DC-plasma nitrided Al–Cu alloy at 673 k with increasing the nitriding duration.

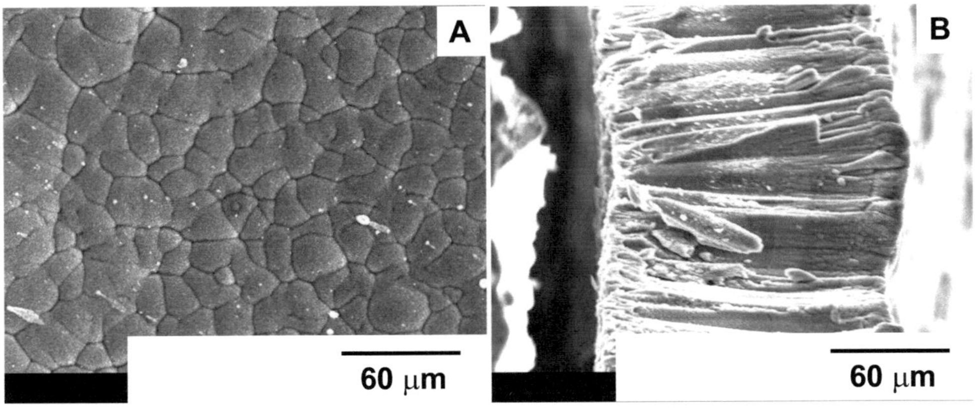

Figure 3.
SEM image on the DC-plasma nitrided Al–Cu alloys at 673 k for 28.8 ks. a) Surface of nitrided layer, and b) cross-section of nitrided layer.

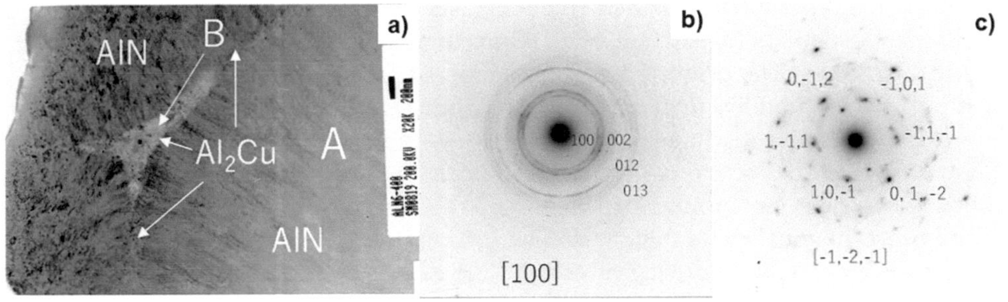

Figure 4.
Coherency between adjacent AlN and Al_2Cu. a) Interfacial boundary B among AlN columns, b) crystallographic structure of AlN in the A-zone, and c) crystallographic structure of AlN and Al_2Cu in the B-zone.

nitriding reaction. The nitrogen diffusion coefficient becomes low in AlN. In addition, no nitrogen is solvable in Al_2Cu. Therefore, interfacial boundaries between the formed AlN and Al_2Cu become an effective diffusion path for nitrogen in this nitriding process. Let us analyze this interfacial boundary between AlN and Al_2Cu in the following.

A crystallographic analysis was conducted to describe the coherency between adjacent Al_2Cu and AlN. As depicted in **Figure 4a**, numerous interfacial zones are observed among AlN columns, where AlN and Al_2Cu are present in adjacency. AlN has a hexagonal structure with a = 0.497 nm and b = 0.497 nm, as shown in **Figure 4b**. Meanwhile, Al_2Cu or θ -phase in the phase diagram of Al–Cu base alloys, has a tetragonal structure with a = 0.606 nm and c = 0.487 nm. For the estimation of the crystallographic coherency between AlN and Al_2Cu, their d-spacing distances were calculated. As demonstrated in **Figure 4c**, in the case of AlN, the d-spacing between {-101} and {-11-1} planes is 0.237 nm, and the d-spacing between {121} and {211} planes is 0.236 nm in the case of Al_2Cu. Both values are in fairly good agreement. This confirms that AlN is synthesized in

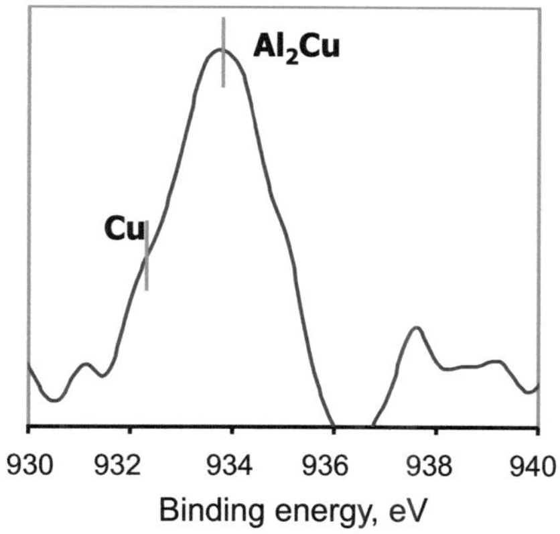

Figure 5.
*XPS analysis of the copper binding state at the b-zone in **Figure 4a**.*

coherency with Al_2Cu. AlN is formed by the reaction between Al_2Cu and diffusing nitrogen solute through the interfacial boundaries; e.g., $Al_2Cu + 2N \rightarrow 2AlN + Cu$. Hence, metallic copper must be present at the interfacial boundaries. Let us perform XPS analysis on the copper binding state at the B-zone.

As shown in **Figure 5**, metallic copper coexists with Al_2Cu; the AlN formation reaction advances in the interfacial zones via $Al_2Cu + 2N \rightarrow 2AlN + Cu$. Remembering that the initial content of Cu is only 6 mass%, a significant amount of Al_2Cu is reproduced in the nitrided layer during the nitriding process. Residual Al_2Cu and Cu in the nitrided columnar structure in **Figure 5** also reveal that the solid-state reaction, represented by $Al_2Cu + 2N$ balu $2AlN + Cu$ and 2Al (in matrix) + Cu balu Al_2Cu takes place during this nitriding process for the formation of AlN and the reproduction of Al_2Cu, respectively. This solid-state reaction becomes a driving process, utilizing Al_2Cu as a catalytic material in plasma nitriding.

3.2 Mechanical characterization

The micro-Vickers hardness testing was used to describe the transients of the hardness depth profile with increasing nitriding duration. **Figure 6** depicts the advancement of hardness depth (h) with increasing nitriding time. It grows linearly with E; e.g., $dh/d\tau$ becomes nearly constant. As stated in [25, 26], E^2 is theoretically or experimentally proportional to the nitriding duration (τ) if the nitriding behavior is mainly governed by nitrogen diffusion. **Figure 6** implies that the fast-rate nitriding process is never simply governed by nitrogen diffusion when it is accommodated by the catalytic reaction of Al_2Cu precipitates.

The thick nitrided layer is a composite material of AlN with Al_2Cu, the volume fraction of which is estimated to be 10 vol% in total. The mixture rule of hardness is used to estimate its hardness from 1,400 HV for pure AlN and 250–350 HV for Al_2Cu [27]. Its measured hardness, 1200 HV in **Figure 7**, is in good agreement with the estimated value of 1200–1300 HV.

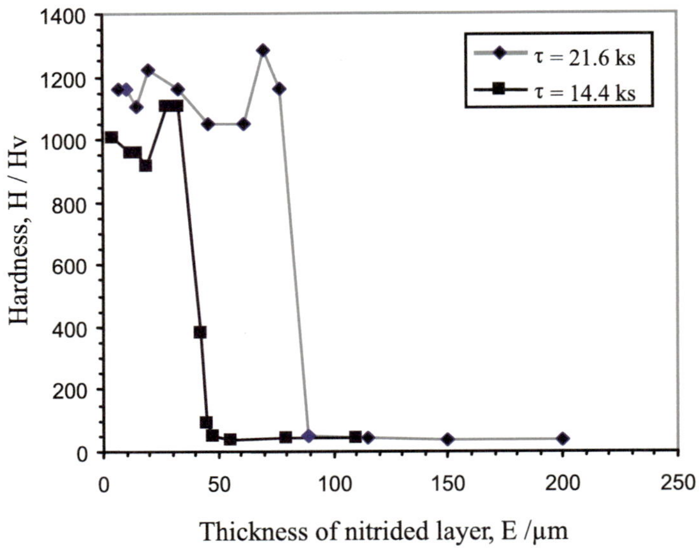

Figure 6.
Transients of hardness depth profile with increasing nitriding time.

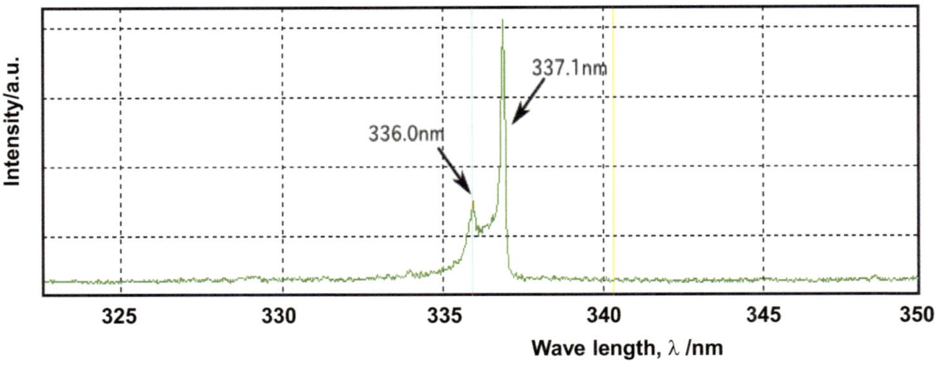

Figure 7.
Emissive light spectrum of the ignited nitrogen–hydrogen plasmas in the narrow wavelength range from 320 nm to 350 nm.

4. RF–DC plasma nitriding of Al–Cu alloys at 673 K

Al–Cu alloys, or AA2011 alloys, were employed to describe the plasma nitriding behavior at 673 K. In a similar manner to DC-plasma at 673 K, the AlN-layer was formed by the catalytic role of Al_2Cu precipitates.

4.1 Plasma diagnosis

The emissive light spectroscopy (PMA-11, Hamamatsu-Photonics Co. Ltd.; Hamamatsu, Japan), as well as the Langmuir probe system (ALP System, Impedance

Co. Ltd., Irland) were instrumented to the present RF–DC plasma nitriding system to make an in situ quantitative diagnosis of the generated nitrogen–hydrogen plasmas. In particular, a high-resolution spectroscopic analyzer with a resolution of 0.1 nm was utilized to detect the NH radicals in the specified wave length range. Furthermore, SPECAIR [28] was used as a simulator of emissive light spectra to theoretically predict the synthesis of NH-radicals in plasma nitriding under the mixture of nitrogen and hydrogen gases.

In this quantitative diagnosis of plasmas, the hydrogen-to-nitrogen gas flow rate was varied as a parameter with significant influence on NH-radical synthesis, as noted in [29, 30]. First, the emissive light spectra from nitrogen-hydrogen plasma were detected and analyzed using high-resolution spectroscopy under the standard experimental conditions outlined in **Table 7**. When scanning the measured spectra across a wider wave length range, the activated nitrogen atom (N^* or N(I)), nitrogen ions (N^+ or N(II)), and N^{2+} (N(III)) were detected to have significant peaks, along with the ionized nitrogen molecule (N_2^+). This synthesis of nitrogen ions is attributed to a series of reactions originating from the parent species N_2^+; e.g., $N_2^+ \rightarrow N^* + N^+$, $N^+ \rightarrow N^{2+} + e$, and $N^+ + e \rightarrow N^*$. Next, the wave length range was narrowed to 290–350 nm. The NH radicals were detected at λ = 336 nm with relatively high intensity, comparable to N_2^+ peak at λ = 337.1 nm, as depicted in **Figure 7**. The ratio of NH-radical peak intensity to N_2^+ peak intensity increases with decreasing hydrogen content; the hydrogen gas flow rate relative to nitrogen gas is expected to be a key parameter in nitriding, which will be discussed later.

In parallel with this experimental diagnosis, the program SPECLAIR was employed to simulate the NH radical spectra, as referenced in Refs. [31, 32]. Among several input parameters for this simulation, the electron temperature (T_e) and its density (N_e) were directly measured using the Langmuir probe [33]; e.g., T_e = 5 eV and $N_e = 4 \times 10^{16}\ m^{-3}$. The experimentally analyzed spectra of NH radicals generated in the nitrogen–hydrogen gas mixture with a ratio of four to one were in good agreement with the theoretically estimated spectra by SPECAIR, as seen in **Figure 8**. This assures that a significant intensity of NH radicals is generated under these plasma conditions with lower hydrogen content. The high electron density reveals that nitrogen ionization and NH-radical formation are enhanced under this plasma nitriding condition. That is, the experimental conditions in **Table 7** were suitable for generating plasmas with the flux of N^*, N^+, N^{2+}, N_2^+, and NH-radicals.

4.2 Plasma nitriding of AA2011 at 723 K and 673 K

AA2011 Al–Cu alloy specimens were employed for RF–DC plasma nitriding at 723 K and 673 K. **Figure 9** depicts the XRD diagram of the nitrided AA2011 disc specimen at 723 K. In addition to the aluminum peaks with the first and second highest intensities, respectively, at 2 θ = 38.5° and 44.8°, typical triplet peaks for aluminum nitride (AlN) were detected at 2 θ = 33.4°, 36.1°, and 38.0°, respectively. This assures that the nitrided layer consists of a composite of synthesized AlN phase and AA2011 matrix. Comparing this XRD diagram to that of DC-plasma nitrided Al–Cu alloy for 14.4 ks, both are in fairly good agreement with each other. This implies that nearly the same nitriding process advances into the depth of the Al–Cu alloy matrix in both plasma nitriding treatments.

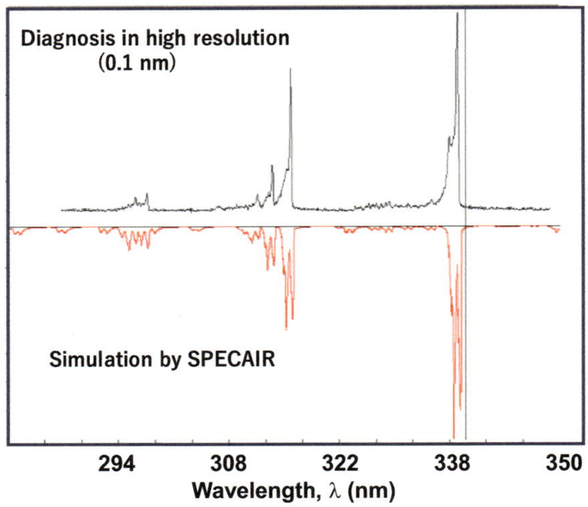

Figure 8.
Comparison of the emissive light spectra between the experimental measurement and the theoretical simulation.

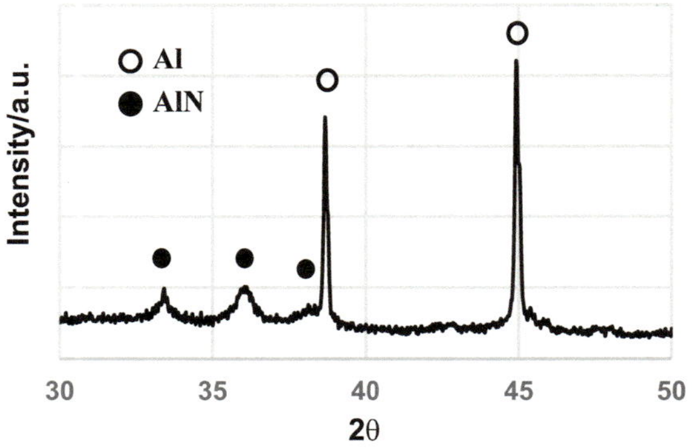

Figure 9.
XRD diagram of RF–DC plasma-nitrided AA2011 specimen at 723 K for 14.4 ks.

The plasma-nitrided AA2011 disc specimen, under the experimental conditions outlined in **Table 7**, is shown in **Figure 10a**. The specimen surface was homogeneously nitrided, resulting in a black coloration. **Figure 10b** depicts its microstructure. The black zone corresponds to the nitrided area, while the white zone represents the original matrix. The area fraction of the nitrided zones relative to the whole surface area is 80%, as shown in **Figure 10b**. Subsequently, its volume fraction (f) is estimated to be approximately 60%. The mechanical properties of this composite microstructure are evaluated through micro-hardness testing in the following sections.

Most of the copper in AA2011 or Al–Cu binary alloy is bound in the form of Al_2Cu as a precipitate. This Al_2Cu first precipitates during hot extrusion and exists in the vicinity of the grain boundary. This original precipitate still coexists with AlN, as discussed in Refs. [18–20]. This Al_2Cu-precipitate also works as a catalyst in the RF–DC plasma nitriding to accommodate the fast-rate nitriding by a series of solid-phase reactions:

Figure 10.
Optical-microscopy and SEM observation of the RF–DC plasma nitrided AA2011 specimen at 723 K for 14.4 ks.

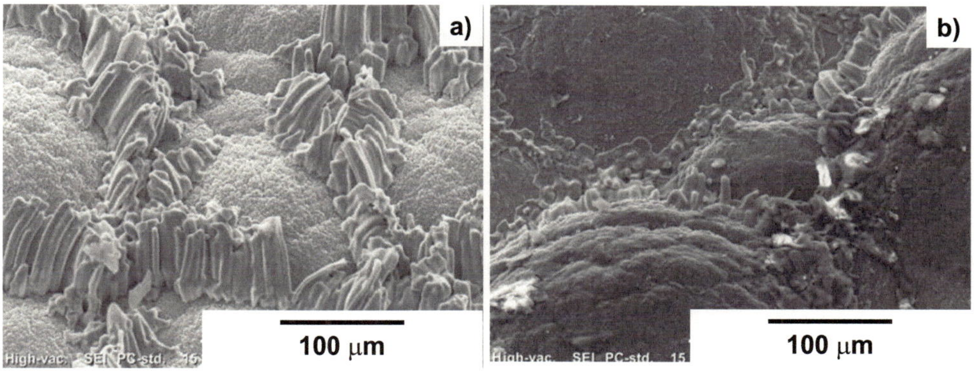

Figure 11.
AlN nucleation and growth processes in the RF–DC plasma nitriding of AA2011 at 673 K. a) SEM image on the AlN nucleation along the grain boundaries at τ = 600 s, and b) SEM image on the AlN growth at τ = 3.6 ks.

$Al_2Cu + 2N \rightarrow 2AlN + Cu$, and $Cu + 2Al \rightarrow Al_2Cu$. Then, an AlN-cluster is expected to nucleate at the grain boundaries, including Al_2Cu precipitates. To describe this AlN-cluster formation, the presputtering temperature and the holding temperature were reduced to 673 K in **Table 7** to prepare the nitrided AA2011 specimens in the early stage of nitriding by controlling the nitriding duration for microstructural analysis.

Figure 11 shows the SEM image of nitrided AA2011 specimen surfaces at 673 K by varying the nitriding duration (τ). When τ = 600 s, the AlN embryos are formed at the grain boundaries, as seen in **Figure 11a**. This proves that both the nitrogen solute isolating from NH radicals and the nitrogen ions react with Al2Cu precipitates and nucleate the AlN embryos at the grain boundaries through the chemical reactions by $Al_2Cu + 2N \rightarrow 2AlN + Cu$ and $Cu + 2Al \rightarrow Al_2Cu$. When τ = 3.6 ks, these embryos coalesce with each other and form a thin aln-layer on the granular surface of AA2011 specimen, as seen in **Figure 10b**. This observation in **Figure 11** reveals that the nitriding process of Al–Cu alloys advances by the nucleation and growth mechanism of AlN-clusters under the intense nitrogen flux from the surface during the RF–DC plasma nitriding.

4.3 Mechanical characterization of RF–DC plasma nitrided AA2011 at 723 K

Variation of surface hardness with increasing weight (W) of the indenter is shown in **Figure 12**. In the case where W = 0.1 N or 10 gf, the surface hardness was measured to be 800 H. This hardness is eight times greater than the original hardness of AA2011 alloy. With increasing weight of the Vickers indenter, the measured surface hardness decreased. According to [34], the matrix hardness influences the measured hardness of materials since the matrix, six times deeper than the indentation depth, is forced into elasto-plastic deformation during indentation for hardness measurement. From this gradual decrease in surface hardness with W, the equivalent depth of the nitrided layer was estimated to be 15 μm when nitriding at 723 K for 14.4 ks.

The nitrided layer becomes a composite of AlN-zones and an Al–Cu alloy matrix, as shown in **Figure 10b**. Assuming that the AlN-zones are homogeneously distributed in the entire nitrided layer by the volume fraction (f), as seen in **Figure 10b**, the linear law in homogenization theory [35] provides that the theoretical hardness (H_T) is estimated by $H_T = (1\text{-}f) \times H_{matrix} + f \times H_{AlN}$, where H_{matrix} is the hardness of AA2011 matrix, and H_{AlN} is the hardness of AlN. Since $H_{matrix} = 100$ HV, $H_{AlN} = 1400$ HV, and f = 0.6, H_T is estimated to be 860 HV. This H_T is in good agreement with the measured surface hardness, which ranges from 800 to 900 HV. This agreement confirms that the nitrided layer should be a composite of AlN and an Al–Cu alloy matrix, with a thickness of 15 μm.

5. RF–DC plasma nitriding of Al–Cu–Mg alloys at 723 K and 623 K

Most of the alloying elements in the aluminum alloys with lower melting temperatures have an influence on the nitriding behavior. AA2014 or Al–Cu–Mg alloy, was employed to describe this influence of Mg on the nitriding of the aluminum matrix. The holding temperature (T_H) was varied to describe the thermal activation effect on the nitriding behavior. Magnesium diffusion is easily activated at T_H = 723 K; however, it is retarded at T_H = 623 K.

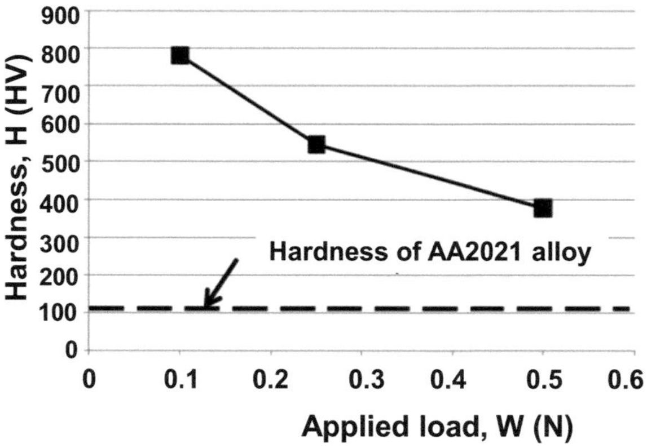

Figure 12.
Variation of surface hardness with increasing the applied load in micro-vickers testing.

5.1 RF–DC plasma nitriding of Al–Cu–Mg alloys at 723 K

AA2014 specimen was also prepared for plasma nitriding at 723 K for 14.4 ks. The nitrided layer thickness was too thin, only to have the surface hardness from 100 to 120 HV. XPS was utilized to investigate this thin film formed on the AA2014 surface.

The chemical binding state of N1s in the vicinity of the nitrided surface is depicted in **Figure 13a**. The binding energy at the highest peak is 397 eV, corresponding to 397.3 eV for the binding energy of N1s in AlN. **Figure 13b** depicts the chemical binding state of Al2p. Since the binding energy of Al2p in metallic aluminum, Al_2O_3 and AlN is 72.7 eV, 74.1 eV, and 74.0 eV, respectively, the aluminum in this thin film is bound in AlN and Al_2O_3. This XPS analysis reveals that AlN-formation is significantly hindered during plasma nitriding at 723 K by the surface oxidation reaction.

XPS, with an etching rate of 1–2 nm/min, was utilized to perform chemical analysis from the surface to the depth. At each etching time, Al2p, Cu2p, Mg2s, O1s, and N1s diagrams were detected to calculate the integrated area of each peak profile and deduce the concentration of each of the five elements. **Figure 14** depicts the variation of Al, Mg, Cu, N, and O concentrations with the etching time in the vicinity of AA2014 alloy surface after plasma nitriding at 723 K under the conditions outlined in **Table 7**. The nitrogen and copper contents were detected at sufficient concentrations to synthesize AlN through the reaction: $Al_2Cu + 2N \rightarrow 2AlN + Cu$, and $Cu + 2Al \rightarrow Al_2Cu$. The magnesium content reaches 3–5 at%, which is much higher than the magnesium composition listed in **Table 6**. The high oxygen content detected in **Figure 14** poses a risk of deteriorating the nitriding reaction due to surface oxidation, as seen in **Figure 13b**. As suggested in [21], magnesium solutes easily diffuse onto the surface through the grain boundaries, blocking the reaction between aluminum and diffusing nitrogen, and assisting the oxidation reaction between aluminum/magnesium and oxygen.

Although AA2011 or Al–Cu binary duralumin, was successfully nitrided under the same conditions as those in **Table 7**, the nitrogen diffusion and its reaction to synthesize AlN were completely retarded by magnesium concentration and surface oxidation during plasma nitriding at 723 K. A higher concentration of magnesium than the original content of 0.5 mass% might be attributed to the intense diffusion of

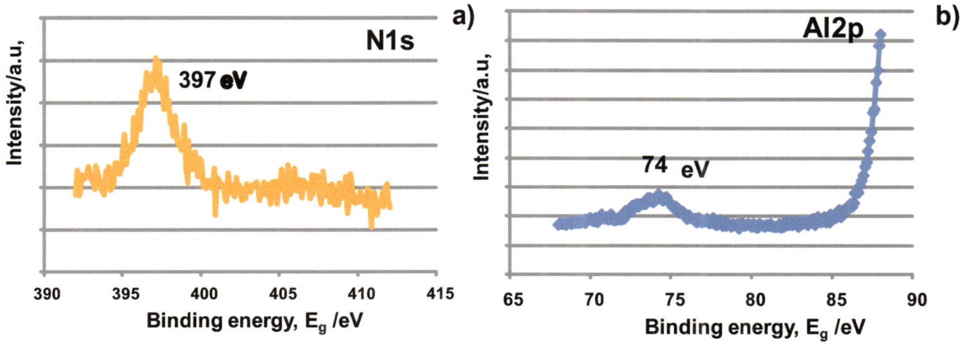

Figure 13.
XPS analysis of the surface of RF–DC plasma-nitrided AA2014 specimen at 723 K. a) chemical binding state of nitrogen, and b) chemical binding state of aluminum.

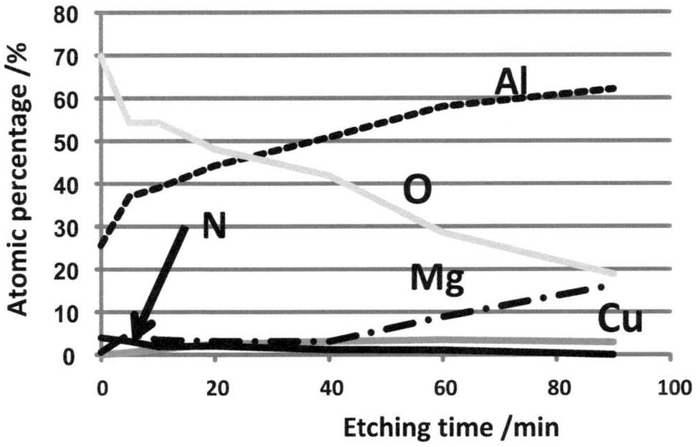

Figure 14.
Variation of the related elements (Al, Cu, Mg, O, and N) to nitriding behavior of AA2014 with the depth from the surface.

magnesium solute from the depth of AA2014 matrix to the surface during nitriding at 723 K (or 450°C). The magnesium solute diffuses to the surface and reacts with oxygen to form magnesium oxides at this holding temperature [36].

5.2 RF–DC plasma nitriding of Al–Cu–Mg alloys at 623 K

Considering that this retardation by Mg-concentration at the surface should be only thermally controlled, the decrease of holding temperature during nitriding becomes a solution to be free from Mg-concentration and related surface oxidation, and to enable plasma nitriding of Al–Cu–Mg alloys. The plasma nitriding experiment was performed at a much lower holding temperature to reduce magnesium diffusion and oxidation. In **Table 7**, the holding temperature during the presputtering and plasma nitriding processes was fixed at 623 K.

Materials characterization. The over view of the plasma-nitrided AA2014 disc specimen at 623 K for 14.4 ks is shown in **Figure 15a**. The entire surface is colored black. Its XRD diagram is shown in **Figure 15b**. Just as analyzed in **Figure 9** for the nitrided AA2011 specimen, this diagram is characterized by two peaks for the

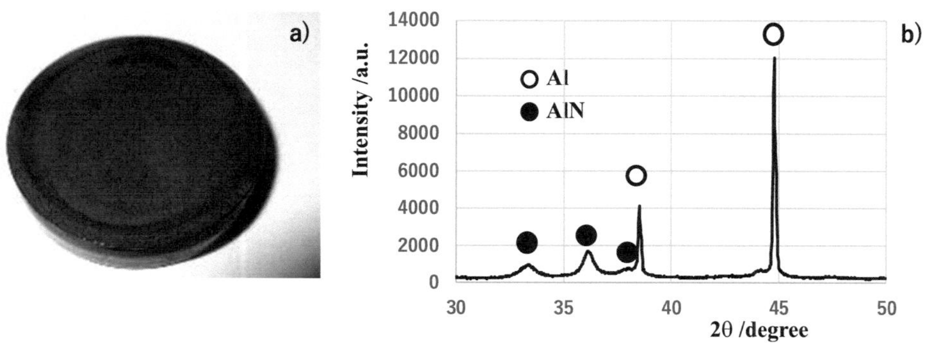

Figure 15.
RF–DC plasma nitrided AA2014 at 623 K. a) Surface structure of nitrided AA24 specimen, and b) its XRD diagram.

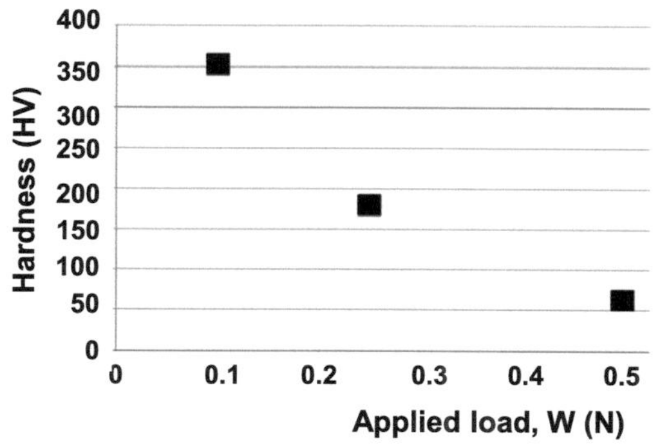

Figure 16.
Variation of the surface hardness with increasing the applied load for RF–DC plasma nitrided AA2014 at 623 K for 7.2ks.

aluminum matrix and triplet peaks for AlN. This proves that AA2014 alloy is plasma nitrided to form the AlN–AA2014 composite layer via the AlN-cluster nucleation and growth mechanism, in a manner similar to the nitriding behavior observed in the AA2011 specimen. Under the lower holding temperature, the magnesium migration through the Al–Cu–Mg alloy matrix is significantly retarded, sustaining the precipitation reaction to synthesize AlN between aluminum in the matrix and the diffusing nitrogen solutes via the total reaction of Al + N → AlN, with the aid of Al_2Cu precipitates.

Mechanical characterization. Under the RF–DC plasma nitriding conditions outlined in **Table** 7, the nitriding duration was reduced to 7.2 ks to describe the surface hardening of AA2014 in the early stage of nitriding. The variation of surface hardness with increasing applied load during micro-Vickers testing is depicted in **Figure 16**. Although the surface hardness was lower than 800 HV for the RF–DC plasma-nitrided A2011 at 723 K (or 450 °C) for 14.4 ks, the A2014 alloy was also plasma-nitrided to achieve a higher hardness of 400 HV compared to the matrix hardness of 100 HV. This confirms that the nitriding process is effective even in the Al–Cu–Mg ternary alloy system, without requiring Mg-concentration at the surface or causing significant oxidation.

5.3 Application of RF–DC plasma nitriding to aluminum alloy parts

Among a wide variety of aluminum automotive parts, a heatsink has become attractive for the application of duralumin alloys. The heatsink part requires high specific strength and high thermal conductivity to promote the cooling capacity of heated power transistors or LEDs (light emitting diodes). The AlN-capped duralumin alloy heatsink is expected to have high thermal conductance and wear resistance. Most aluminum automotive parts have been made of die-cast or gratitude-cast alloys. In the present experiment, the die-cast AA2011 heat sink was first fabricated to have fin alignment for heat transfer, where the fin thickness was 1 mm, the fin height was 10 mm, and its distancing pitch between adjacent fins was 2 mm, as shown in **Figure 17a**.

Figure 17 compares the plasma-nitrided AA2011 heatsink part at 673 K for 14.4 ks under the conditions in **Table 6** to the bare die-cast AA2011 heatsink. The head and

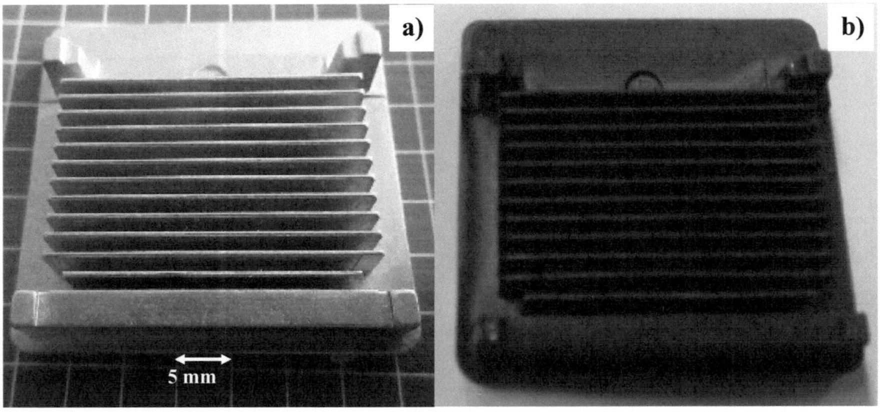

Figure 17.
A die-cast AA2011 aluminum-alloy sink. a) as-die-cast AA2011 heat sink, and b) rf–dc-plasma nitrided AA2011 heat sink at 673 k for 14.4 ks.

side surfaces of the fins are finely nitrided to achieve a surface hardness of 800 HV. This AlN-capping onto AA2011 die-cast heatsink demonstrates the high feasibility of the RF–DC high-density plasma nitriding process in industrial applications.

6. Discussion

In the DC- and RF–DC plasma nitriding of Al-Cu alloy or AA2011 at 673 K, an AlN-rich surface layer is formed as a composite of AlN and Al_2Cu within the aluminum alloy matrix. This Al_2Cu plays a catalytic role in driving the reaction via $Al_2Cu + 2\,N \rightarrow 2AlN + Cu$ and $2Al + Cu \rightarrow Al_2Cu$. The nitriding behavior, facilitated by this catalytic reaction, is characterized by the rapid formation of the AlN layer through the overall reaction of Al (in matrix) + N (diffusing nitrogen) → AlN. The isolated nitrogen solutes from NH radicals and nitrogen ions, in addition to activated nitrogen atoms, do not diffuse through AlN but rather through the interfacial boundaries between AlN-clusters and catalytic Al_2Cu precipitates. This Al_2Cu accommodating both the catalytic reactions and the nitrogen solute diffusion network, drives the growth of a thick AlN-layer at a faster rate than that observed in the normal nitriding process [9–14].

DC-nitriding process of Al-Cu alloys is designed to improve the yield of NH radicals by using a mixture of nitrogen–hydrogen gas with a flow rate ratio of 75% for nitrogen and 25% for hydrogen. This NH radical promptly decomposes into activated nitrogen and hydrogen atoms on the surface of the Al-Cu alloy. The nitrogen atom impinges into the matrix as a reactive solute to Al_2Cu. The hydrogen atom recombines to form a hydrogen molecule in the gas phase. RF–DC plasma is ignited under the mixture gas with 80% nitrogen and 20% hydrogen. The yield ratio of the NH-radical population to N_2^+ population reaches its maximum, driving the nitriding process through chemical nitriding via NH-radical attack on the Al–Cu alloy, together with mechanical nitriding via N_2^+ bombardment and nitrogen penetration.

Let us discuss the effect of hydrogen flow rate on this yield ratio in the RF–DC plasma nitriding. **Figure 18** depicts the variation of maximum surface hardness with an increasing H_2/N_2 ratio in the mixture gas. When the ratio is less than 20%, or the

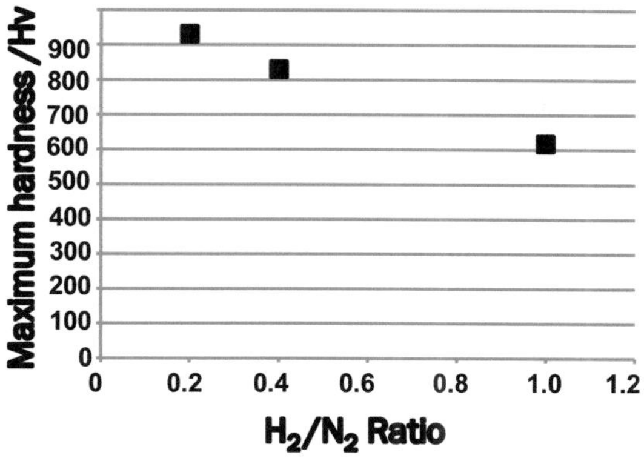

Figure 18.
Variation of measured maximum surface hardness with increasing H_2/N_2 ratio in the mixture gas.

N_2 to H_2 ratio is more than 4 to 1, the maximum surface hardness becomes nearly the same, ranging from 950 HV to 1000 HV. A relatively thick AlN-layer is yielded in this low-ratio regime. On the other hand, the maximum surface hardness monotonously decreases with the H_2/N_2 ratio in **Figure 18**. This is because of the so-called hydrogen quenching effect [29–31]. This reveals that efficient nitriding could be advanced by RF–DC plasmas under the optimum control of gas supply.

XRD diagrams in **Figures 2** and **9** reveal that AlN formation by nitriding for 14.4 ks becomes equivalent to each other for DC- and RF–DC plasma nitriding.

Magnesium in AA2014 or Al–Cu–Mg alloys poses a risk of deteriorating the nitriding behavior due to its concentration at the surface via migration from the depth and its oxidation at the surface when the holding temperature exceeds 700 K. This deterioration can be suppressed by lowering the holding temperature to 623 K. This reveals that magnesium migration to the surface and oxidation is a thermally activated process; the original AlN formation process can be reproduced by plasma nitriding at lower temperatures. During plasma nitriding, the holding temperature must be lowered as much as possible to suppress the surface diffusion of low melting-point metal elements in the aluminum alloy matrix.

Let us discuss how to effectively lower the holding temperature while preserving the nitriding capacity to form a thick AlN-layer in various aluminum alloys.

As stated in [37, 38], austenitic and martensitic stainless steels were successfully RF–DC plasma nitrided by massive nitrogen supersaturation (MNS). A thick nitrogen-supersaturated layer with a thickness of 60 μm was formed without any nitride precipitates by RF–DC plasma nitriding at 673 K. Due to this MNS, the hardness was enhanced to 1400 HV. As analyzed in [39, 40], the homogeneous MNSed layer was also formed by RF–DC plasma nitriding at 623 K for fine-grained AISI316 and AISI316L stainless steels. This success of low-temperature RF–DC plasma nitriding reveals that massive nitrogen supersaturation provides a way to form a thick solid-solution-hardened layer even at lower temperatures than 600 K without the use of intermetallic compound (IMC) precipitates.

As a typical IMC precipitate in Al–Cu and Al–Cu–Mg alloys, Al_2Cu plays a catalytic role in initiating the exothermic reactions to form the AlN layers and sustaining the chained reaction for the growth of the AlN-layer. In the presence of an ignitor to drive the reaction between the diffusing nitrogen solute and the fine IMC-precipitate (IMP), such as Al_2Cu precipitate to drive 2 N + Al_2Cu → 2AlN + Cu can be initiated. The subsequent reaction via 2Al + Cu → Al_2Cu can be sustained even at low holding temperatures. That is, Al_xM for $x > 1$ works as an ignitor to onset the reaction of xN + Al_xM → xAlN + M; then, xAl + M → Al_xM recovery reaction should take place at lower holding temperatures. In the case of low-temperature plasma nitriding for other series of aluminum alloys such as AA5000, AA6000, or AA7,000, alternative IMCs must be designed to act as catalysts to sustain the exothermic reactions even at temperatures lower than 600 K.

Heat sink made AA2011 alloy was successfully nitrided at 673 K for 14.4 ks, achieving a hardness of more than 800 HV. This homogeneous nitriding in the heat sink fins ensures that duralumin parts and members with small holes and asperities are effectively nitrided using the present approach. This plasma nitriding method offers a new way of surface treatment for various duralumin alloy parts and members. Toward efficient plasma nitriding of both Al–Cu and Al–Cu–Mg alloys in commercial grades, the present RF–DC plasma system can be further improved to enable low-temperature nitriding through high densification of ions and radicals under relatively low electron temperatures.

7. Conclusion

DC- and RF–DC plasma nitriding of Al–Cu alloys at 673 K proves that a thick AlN-rich composite layer is rapidly formed as a hardened surface region. In this nitriding process, Al_2Cu plays a catalytic role in initiating and sustaining the exothermic reactions. Fine precipitates in Al-Cu alloys result in the homogeneous formation of the AlN-rich composite layer. In the case of plasma nitriding the Al–Cu–Mg alloys, the holding temperature must be sufficiently low to retard the magnesium solute diffusion to the surface and to minimize the disturbance to the nitriding behavior caused by surface oxidation.
Without deterioration due to magnesium diffusion, the AlN-rich layer can be formed for 14.4 ks even at 623 K in a similar manner to DC-plasma nitriding at 673 K.

An A2011 heat sink was prepared by die-casting and RF-DC plasma nitrided at 673 K for 14.4 ks to form a homogeneous, thick AlN-layer. Its high hardness and thermal conductivity make it suitable for heat spreading in power transistors, CPUs, and telecommunication relay stations.

Two approaches toward a lower plasma nitriding process are proposed for further development: massive nitrogen supersaturation into aluminum alloys, and intermetallic precipitate accommodation to AlN-layer formation. In the former, a massively nitrogen-supersaturated layer is formed in aluminum alloys using the fine grain boundary network as a nitrogen diffusion path to sustain the nitriding behavior without the formation of nitrides, including AlN. In the latter, intermetallic compound precipitates (IMP) are designed to work as a catalyst to accommodate the AlN-layer into the aluminum alloy matrix at a lower holding temperature. These more advanced nitriding processes enhance the reusability for upgrading recycled aluminum and aluminum alloys.

Acknowledgments

Authors would like to express their gratitude to Prof. Patama (PhD student at the University of Tokyo; now Associate Professor, Chulalongkorn University, Thailand) and S.–I. Kurozimi (Nano-Coat Lab., LLC.) for their help in experiments.

Conflict of Interest

The authors declare no conflict of interest.

Author details

Tatsuhiko Aizawa[1*], Hiroki Nakata[2] and Takeshi Nasu[3]

1 Surface Engineering Design Laboratory, Shibaura Institute of Technology, Tokyo, Japan

2 Hiroki Nakata, Ebinax, Co., Ltd., Tokyo, Japan

3 Takeshi Nasu, Ebinax, Co., Ltd., Tokyo, Japan

*Address all correspondence to: taizawa@sic.shibaura-it.ac.jp

References

[1] Zanzoni A, Montecchi-Palazzi L, Quondam MX. Mint, A molecular interaction database. FEBS Letters. 2002;**513**:135–140. DOI: 10.1016/s0014-5793(01)03293-8

[2] Al-Alimi S, Yusuf NK, Ghaleb AM, Lajis MA, Shamsudin S, Zhou W, Altharan YM, Abdulwahab HS, Saif Y, Didane DH, Ikhwan STT, Adam A. Recycling aluminium for sustainable development: A review of different processing technologies in green manufacturing. Results in Engineering. 2024;**23**:102566

[3] Ardelean M, Lascău S, Ardelean ED, Josan A. Surface treatments for aluminium alloys. IOP Conference Series: Materials Science and Engineering. 2018;**294**1:012042

[4] Quebbou Z, Chafi M, Omari LEH. Corrosion resistance of 5005 aluminum alloy by anodizing treatment in a mixture of phosphoric and boric acids. Materials Today: Proceedings. 2021;**37** (3):3854–3859

[5] Somers MAJ, Christiansen TL. Nitriding of Steels. Encyclopedia of Materials: Metals and Alloys. 2022;**2**:173–189

[6] Zhou YL, Xia F, Xie -J, Peng HP, Wang JH, Li ZW. A review—effect of accelerating methods on gas nitriding: Accelerating mechanism, nitriding behavior, and techno-economic analysis. Coatings. 2023;**13**(11):1846. DOI: 10.3390/coatings13111846

[7] Balikci E, Yaman O. Investigation on liquid bath nitriding of selected steels. Surface Engineering. 2011;**27** (8):609–615. DOI: 10.1179/1743294411Y.0000000034

[8] Reddy CAK, Srinivasan T, Venkatesh B. Effect of plasma nitriding on M50 NiL steel – A review. Materials Today: Proceedings. 2022;**52**(3): 1073–1077. DOI: 10.1016/j.matpr.2021.10.494

[9] Shim YK, Kim YK, Lee KH, Han S. The properties of AlN prepared by plasma nitriding and plasma source ion implantation techniques. Surface and Coatings Technology. 2000;**131**:345–349

[10] Chakraborty J, Mukherjee S, Raole PM, John PI. Feasibility study of aluminium nitride formation by nitrogen plasma source ion implantation on aluminium. Materials Science and Engineering A. 2001;**304-306**:910–913

[11] Czerwiec T, Michel H, Bergman E. Low-pressure, high-density plasma nitriding: Mechanisms, technology and results. Surface and Coatings Technology. 1998;**108-109**:182–190

[12] Telbizova T, Parascandola S, Prokert F, Barradas NP, Richter E, Molles W. Ion nitriding of Al: Growth kinetics and characterisation of the nitride layer. Surface and Coatings Technology. 2001;**142–144**:1028–1033

[13] Arai T, Fujita H, Tachikawa H. Proc. Conf. Ion Nitriding. 1984, 37–41

[14] Quast M, Mayr P, Stock H-R, Podlesal H, Wielage B. In situ and ex situ examination of plasma-assisted nitriding of aluminium alloys. Surface and Coatings Technology. 2001;**135**:238–249

[15] Zhao C, Li CX, Dong H, Bell T. Study on the active screen plasma nitriding and its nitriding mechanism. Surface and Coatings Technology. 2006;**201**(6):2320–2325

[16] Taweesub K, Visuttipitukul P, Tungkasmita S, Paosawatyanyong B. Nitridation of Al–6%Cu alloy by RF plasma process. Surface and Coatings Technology. 2010;**204**:3091–3095

[17] Hara J. Nitriding system. Japan Patent. P2004-190055A, 2002

[18] Visuttipitukul P, Aizawa T. Advanced plasma nitriding for aluminum and aluminum alloys. Materials Transactions. 2003;**44** (12):2695–2700

[19] Vissutipitukul P, Aizawa T. Wear of plasma-nitrided aluminum alloys. Wear. 2005;**259**:482–489

[20] Visuttipitukul P, Aizawa T. Plasma nitriding design for aluminium and aluminium alloys. Journal of Surface Engineering. 2006;**22**:187–195

[21] NDK, Ion Nitriding. https://www.ndkinc.co.jp/en/equipment/plasma.html. (Retrieved at 2025/7/29)

[22] Takeda H, Kaneko H, Onodera S. Radical nitride treatment method. Denki-Seik. 2004;**75**(4):277–281

[23] Aizawa T, Rosadi I, Yunata E. High density RF-DC plasma nitriding under optimized conditions by plasma-diagnosis. Journal of Applied Sciences. 2022;**12**(3706):1–12

[24] Aizawa T, Yoshino T, Fukuda T. High density RF/DC low temperature plasma nitriding of Al-Cu alloys. In: Proc. 18th ICAA Conference 2024; p. 123

[25] Granito N. Surface Material Design and Control in Steels for Automotive Parts by Plasma Nitriding Techniques. PhD Thesis, University of Tokyo 2003

[26] Aizawa T, Sugita Y. High density RF-DC plasma nitriding of steels for die and mold technologies. Res. Rep. SIT. 2013;**57**(1):1–10

[27] Alam SN, Shrivastava P, Panda D, Gunale B, Susmitha K, Pola P. Synthesis of Al2Cu intermetallic compound by mechanical alloying. Materials Today Communications. 2022;**31**:103267. DOI: 10.1016/j.mtcomm.2022.103267

[28] Maher I, Boulos MI, Fauchais PL, Pfender E. Plasma diagnostics, optical emission and absorption spectroscopy. In: Handbook of Thermal Plasmas. Cham: Springer; 2023, p. 963–1023

[29] Rosadi I, Djoko D, Sakti SP, Yunata E, Aizawa T. Plasma diagnosis on the mixture gas plasma state for nitriding process. In: Proceedings of the 12th SEATUC Symposium (Yyakarta, Indonesia; 2018/3/3). 2018, p. 131–136

[30] Sharma MK, Saikia BK, Bujarbarua S. Optical emission spectroscopy of DC pulsed plasmas used for steel nitriding. Surface and Coatings Technology. 2008;**203**:229–233

[31] Kyo M, Takeguchi Y, Uesugi Y, Tanaka Y, Masuzaki S. Suppression of carbon dust formation by nitrogen injection into hydrogen plasmas in detached plasma conditions. Plasma and Fusion Research. 2010;**5**(004):1–3

[32] Aizawa T, Sugita Y. High density plasma nitriding of tool and die steels. In: Proc. 6th South East Asian Technical University Consortium Symposium. 2012, p. 134–137

[33] Santjojo D, Itsiroyah, Aizawa T. Dynamic of nitrogen and hydrogen species in a high-rate plasma nitriding of martensitic stainless steels. In: Proc. 9th SEATUC Conference. 2015, July, Thai. CD-ROM

[34] Umbrello D, Hua J, Shivpuri R.Hardness-based flow stress and fracture models for numerical simulation of hard machining AISI52100 bearing steel. Materials Science and Engineering A. 2004;**374** (1–2):90–100

[35] N-Nasser MHM. Micromechanics: Overall Properties of Heterogeneous Materials. North Holland; 1997

[36] Holun KJ, Matienzo LJ. Magnesium diffusion in several aluminum alloys. Applications of Surface Science. 1981;**9** (1):22–38

[37] Aizawa T. Low Temperature Plasma Nitriding of Austenitic Stainless Steels. Chapter 3 in Stainless Steels and Alloys. London, UK: IntechOpen; 2019. p. 31–50

[38] Aizawa T, Shiratori T, Komatsu T. Integrated manufacturing of fine-grained stainless steels for industries and medicals. In: Sharma A, Duriagina Z, Kumar S, editors. Chapter 1 in Engineering Steels and High Entropy-Alloys. London, UK: IntechOpen; 2020. p. 3–26

[39] Farghali A, Aizawa T. Nitrogen supersaturation process in the AISI420 martensitic stainless steels by low temperature plasma nitriding. ISIJ International. 2018;**58**(3):401–407

[40] Aizawa T, Yoshino T, Shiratori T, Yoshihara S-I. Grain size effect on the nitrogen SuperSaturation process into AISI316 at 623 K. ISIJ International. 2019;**59**(10):1886–1892

Section 4

Corrosion Science and Protection Strategies

Chapter 5

Corrosion Mechanisms and Protective Strategies in Aluminum Alloys

Amlan Das and Shashanka Rajendrachari

Abstract

While aluminum alloys are practically employed in aerospace, automotive, marine, and structural fields for their excellent strength-to-weight ratio and easy workability, corrosion generally occurs in them. Corrosion, particularly in these alloys, is caused by micro-galvanic interaction between local anodes and cathodes created by intermetallic particles and the surrounding matrix. This manifests in various forms of corrosion, such as pitting, intergranular corrosion, selective dissolution, trenching, exfoliation, and surface grain etch-out. The degradation progresses further in conjunction with alloy composition, thermomechanical history, the type of environment the alloy is exposed to, and grain energy linked with defect densities. Studies have primarily focused on understanding the electrochemical nature of corrosion and have indicated that the protective passive layer of aluminum oxide (Al_2O_3) is vital to resisting corrosion, although its efficacy can be undermined under harsh conditions like chloride-rich environments. Techniques to protect aluminum alloys include alloying modifications, surface treatments such as anodizing and conversion coatings, cathodic protection, and corrosion inhibitors. Still, there are issues regarding the balance of mechanical properties versus corrosion resistance, environmentally friendly protective coatings, and a detailed understanding of localized corrosion on the microscale. Therefore, multiscale modeling, advanced characterization procedures, and sustainable material solutions are underway to ensure aluminum alloy components exhibit enhanced durability and longer service life in harsh environments.

Keywords: aluminum, corrosion mechanism, intergranular corrosion, micro-galvanic coupling, intermetallic particles

1. Introduction

Aluminum, once a precious metal until the late 1980s is currently a mainstream element in the metal industry. In the mid-1990s, its production expanded owing to the effects of world wars, and since then, its applications have diversified rapidly into consumer goods, transportation, construction, and packaging. The 1xxx series is made from 99% pure aluminum. It has high electrical and thermal conductivity, good corrosion resistance, and low strength, thus meeting the requirements for electrical transmission lines, chemical equipment, and food packaging. The 2xxx series aluminum alloys with copper (magnesium as a secondary addition in some cases) give a high strength-to-weight ratio and good fatigue resistance. They are generally less corrosion-resistant and are commonly utilized in aerospace applications such as aircraft fuselages. The 3xxx series with manganese has slightly higher strength and corrosion resistance and is typically used in beverage cans, cooking utensils, and HVAC systems. The 4xxx series with silicon decreases the melting point and provides fluidity, where welding wires, brazing materials, and automotive engine components are applications. The 5xxx series aluminum alloys with magnesium provide high strength and good corrosion resistance in marine environments and are used in shipbuilding and cryogenic tanks. The 6xxx series, a combination of magnesium and silicon to form Mg_2Si, is heat-treatable, extrudable, and corrosion-resistant; thus, it's well-suited for structural frames, architectural, and automotive applications. The 7xxx series is zinc-based (including magnesium and copper), allowing the alloys to achieve the very highest strengths that are really required in aerospace structures like aircraft wings and military armor. The 8xxx series alloys use lithium and some other elements to reduce the density and enhance stiffness; thus, they're used in very advanced aerospace and automotive parts. Casting alloys (e.g., A356, A380) use a high content of silicon for fluidity and dimensional stability because of the complexity involved in shaping engine blocks and industrial parts. Despite their desirable strength-to-weight ratio and great formability, aluminum alloys undergo corrosion by different mechanisms like pitting, intergranular corrosion, exfoliation, and stress-corrosion cracking (SCC), all essentially because of electrochemical heterogeneities. Several decades of research, starting from Pourbaix diagrams laying down passive regions and Evans conducting one of the first electrochemical studies, describe microstructural features by which intermetallic particles initiate pitting by local galvanic action, while anodic grain boundary precipitates (e.g., Cu-depleted zones in 2xxx alloys, Mg-rich zones in 5xxx, or η-phase ($MgZn_2$) in 7xxx) cause intergranular corrosion. Researchers have reviewed in detail pitting mechanisms and inhibitions along with the development of a mathematical model for pit growth kinetics. Studies on high-strength 7xxx and 2xxx alloys with elongated grains have shown that the phenomenon of exfoliation stress corrosion has a link to intergranular corrosion propagating parallel to the rolling direction. Being an intricate interaction of stress, environment, and microstructure, SCC has been studied extensively by means of fracture mechanics and also surface analysis techniques such as TEM and APT. It has also been observed that the presence of chloride concentration, temperature, pH, or an oxidizing agent can change the impact of each environmental variable considered in corrosion.

Protection techniques have evolved significantly over the past decades owing to the research carried out in the areas of coatings, inhibitors, alloy design, and surface modification. Chromate conversion coatings used to be quite prevalent due to their unmatched self-healing properties, but their toxicity levels forced the research community to find alternatives. These alternatives include trivalent chromium processes, rare-earth-based coatings, sol-gels, layered double hydroxides for inhibitor storage, plasma electrolytic oxidation, and anodizing systems often coupled with nanoparticle sealants. Inhibitor studies are focused on greener alternatives such as organic inhibitors, rare-earth cations, and synergistic mixtures, which are studied through electrochemical and surface analysis techniques. Alloy design strategies favor microstructural control to decrease the size and distribution of inter-metallic phases, heat treatment designs to minimize anodic grain boundary pathways, and the development of low-Cu or Sc/Zr micro-alloys for improved corrosion resistance. Marine environments prefer the use of cathodic protection along with the avoidance of crevices and dissimilar metal contacts, as per modern design practices. Research relevant to the protection of aluminum alloys could be intensively directed into various areas, such as the development of enhanced protection systems, mainly smart coatings that would perform better than the well-known chromate-based systems. These coatings can be designed with features such as stimulus-dependent inhibitor release and self-healing ability, based on the use of nanocontainers and stimuli-responsive polymers. Another significant area is that of multi-functional coatings, which protect against corrosion while offering additional properties such as anti-fouling, de-icing, or wear resistance, thereby increasing the all-around functional life of materials in harsh environmental conditions. Computational and artificial intelligence tools have recently been highly promoted, with techniques such as DFT, phase-field modeling, or FEM being implemented alongside machine learning to predict corrosion initiation and propagation, thereby accelerating the discovery of new alloys and coatings. Corrosion phenomena can also be studied in geometrically complex environments, taking into account dynamic and aggressive conditions such as erosion-corrosion, tribocorrosion, and exposure to biofuels. A strong emphasis can be placed on gaining a fundamental understanding of corrosion mechanisms by using highly advanced in-situ methods, including liquid cell transmission electron microscopy, synchrotron-based X-ray tomography, and scanning electrochemical microscopy. These methods allow for nanoscale, real-time visualization of localized corrosion occurrences. Finally, toward the pursuit of sustainability, efforts can be made to support the circular economy by enhancing corrosion resistance in recycled alloys, which typically contain more impurities.

Corrosion-protection strategies for aluminum alloys, with respect to recently reported literature, as summarized in **Table 1** include multifunctional coatings, ecological inhibitor systems, and AI-assisted inhibitor design. Epoxy and composite coatings amended with functionalized silicon dioxide have increased charge transfer resistance and reduced corrosion rates by 97%, while inhibition efficiency by nanocontainer systems from mesoporous silicon dioxide or metal–organic frameworks reaches 87% and maintains it. Hybrid coatings consist of sol–gel films or plasma electrolytic oxidation plus azole inhibitors, providing impedance values 36 times higher than those of bare alloys. The works also focus on the use of natural substances such as plant extracts and oils as adsorption-based inhibitors with variable efficiencies. An online machine-learning tool predicts the performance of inhibitors with a balanced accuracy of 74% by considering molecular descriptors and pH effects. Other studies describe aluminum alloys such as AA2024, 7,075, $AlMg_3$, etc., while emphasizing strategies that combine barrier protection with active, self-healing release for enhanced corrosion resistance.

2. Corrosion mechanism and influencing factors

Corrosion is basically an electrochemical process involving redox reactions, in which an oxidation process on the metal (acting as the anode) involves the loss of electrons to maintain an associated reduction at the cathode. Despite being highly reactive, aluminum exhibits excellent resistance to corrosion in most atmospheric

Category	System	Mechanisms	Methods	Key performance	Novelty
Multifunctional coating	$Ti_3C_2T_x$ MXene + DTMS silane, multilayer	Super-barrier + hydrophobic + mechanical robustness; interfacial coupling via silane; MXene plates as a tortuous-path barrier	EIS, salt-spray, contact angle, adhesion	R_{ct} ↑ by ~ 2 orders versus bare; 480 h neutral salt spray without blistering; contact angle ≈ 118°; adhesion 5B (ASTM D3359)	Clear demonstration of multilayer architecture tuning, barrier + wetting, while retaining adhesion
Multifunctional coating	PU binder + GO (oxygenated & barrier) + BAMS-grafted VN nanoparticles	Synergistic barrier (GO) + catalytic/passivation (VN) + mechanical strengthening	EIS, salt-spray (ASTM B117), microscopy		Cr-free; VN/GO loads optimized to avoid nanoparticle leaching
Inhibitor technology/review	Li_2CO_3, $Li_2C_2O_4$, $LiNO_3$, etc., as leachable species in organic, sol-gel, and conversion layers	Cathodic inhibition + formation of protective Li-aluminate/Al$(OH)_3$; self-healing via pH-triggered Li^+ release	Survey of EIS, SVET, and NSS across literature	Reports multiple systems achieving durable self-healing on AA2024-T3; codifies pitfalls (alkalinity management, leach rate control)	State-of-the-art synthesis/embedding routes and performance envelope; cites Visser et al. (2016) for prototypical coatings
Multifunctional coating (classic within Li-track)	Organic coatings + Li_2CO_3/$Li_2C_2O_4$ reservoirs	Leachable lithium-based self-healing	EIS/SVET	Benchmark evidence for active healing around defects; a foundation for later Li systems	Canonical reference for Li-salt-enabled healing; often replicated or extended
AI/virtual screening	ML platform for structure-based virtual screening of AI inhibitors	Predict inhibitor efficacy; guide eco-friendly selections; map to coating/nanocontainer integration	Curated dataset + supervised ML; case studies & validation	Rapid down-selection of candidates prior to lab work; the article argues it is a significant accelerator for "green" inhibitor discovery	First purpose-built ML platform focused on AI inhibitor virtual screening
AI/predictive materials	Data-driven models (RFR, GPR) predicting E_corr and corrosion rate from composition & thermo-phys. features	Composition-to-corrosion mapping; feature importance (Cu, Zn, etc.)	Potentiodynamic + immersion + ML	Accurate E_corr and rate prediction demonstrates transferable features (specific heat, latent heat, p-electrons)	First systematic ML linking alloy features to corrosion metrics across many 7xxx casts
AI/datasets for coatings/inhibitors	Multidimensional, time-dependent electrochemical database (~LPR, EIS, PDP across ~ 80 molecules)	Data standardization to train ML predicting inhibitor/coating behavior	High-throughput LPR/EIS/PDP + ML best practices	Establishes data quality/control protocols; shows routes to predictive ML on AA2024-T3 inhibitor systems	Critical "Infrastructure" Paper for ML-Ready Corrosion Datasets on Al

Category	System	Mechanisms	Methods	Key performance	Novelty
Active coating	PEO layer + azoles (1,2,4-triazole, BTA) impregnation; PVDF sealing	Multistage hybrid: barrier + self-healing + polymer sealing	EIS, morphology, long-term tests	Demonstrates self-healing azole reservoirs within porous PEO, improved durability after PVDF seal	Representative hybrid active design aligns with Li-review trends toward multi-layer active systems
Active coating	Epoxy + mesoporous SiO_2 nanocontainers loaded with BTA + MBT	Dual-inhibitor synergistic release	EIS, release kinetics	Extended low-frequency impedance retention; better defect tolerance versus single inhibitor	Early proof of dual-cargo synergy; widely cited in 2023–2024 designs
Active coating	Epoxy + Co-MOF@MBT nanocontainers	Stimuli-responsive MOF; inhibitor on-demand	EIS, leaching tests	Higher R_{ct} retention after scribe vs. neat epoxy; responsive MBT release	Emerging MOF-carrier approach for Al epoxy systems
Barrier coating	Epoxy-hardener + functionalized SiO_2 micro-/nano-fillers	Barrier densification; percolation-driven tortuosity	EIS, salt-spray, mechanical	Optimal SiO_2 load boosts	Cr-free
Active sol-gel	Hybrid sol-gel + halloysite nanotubes (HNTs) loaded with MBT; LbL polyelectrolyte caps	pH-triggered inhibitor release; defect-responsive	EIS/SVET exemplars	Classic high	MBT toxicity; HNT generally benign
Multifunctional composite	SiO_2–h–BN(NH_2) nanosheets + PAA-MBT; anodic alumina carrier	Barrier (nanosheets) + hydrophobicity + active-release (PAA-MBT)	EIS, mechanical, wetting	Enhanced mechanics + low water uptake; sustained impedance	Multi-layered, multifunctional design (barrier + active)
Green inhibitors	Plant extracts, gums, drugs, oils	Adsorption (often mixed-type); competitive with azoles in mild chloride	Compiles WL/ EIS/DFT	Highlights biomass-based adsorbates; mechanisms via π-electrons, heteroatoms	Serves as a sourcebook to pair with ML screens for eco-friendly candidates

Table 1.
Progress in aluminum protection techniques [1–7].

and chemical environments. This extraordinary corrosion resistance is due to the spontaneous formation of a thin, dense, and highly adherent layer of aluminum oxide on its surface. The oxide layer develops within seconds to about 1 nm thickness when the metal surface is exposed to either air or moisture, and also possesses the ability to reform immediately when damaged due to its self-healing nature. In general, the nature of corrosion to either protect or degrade aluminum will be dictated by environmental factors, such as temperature, humidity, and immersion in aqueous media, impurities in the system, or alloying elements. The oxide layer protects the system from corrosion: (1) by presenting a surface resistant to dissolution and (2) by being an electrical insulator and preventing the flow of electrons from the oxidizing metal surface to the electrolyte. The overall anodic reaction for aluminum corrosion is:

$$Al \rightarrow Al^{3+} + 3e^-$$

The released electrons participate in cathodic reactions, depending on the environment. In neutral or alkaline aqueous environments with dissolved oxygen, the cathodic reaction is typically the reduction of oxygen:

$$O_2 + 2H_2O + 4e^- \rightarrow 4OH^-$$

Alternatively, in acidic environments, hydrogen ion reduction may occur:

$$2H^+ + 2e^- \rightarrow H_2 \uparrow$$

The aluminum ions (Al^{3+}) produced during oxidation quickly hydrolyze and precipitate as aluminum hydroxide, which further converts to aluminum oxide, reinforcing the passive layer:

$$2Al^{3+} + 3H_2O \rightarrow Al_2O_3 + 6H^+$$

The underlying mechanism driving almost all corrosion modes in aluminum alloys is an electrochemical reaction between localized cathodes and anodes in the aluminum matrix. Such interaction leads to several corrosion types, including pitting corrosion, selective dissolution, trenching, etching out of intermetallic particles, intergranular attack, and exfoliation corrosion. Furthermore, the process of surface and sub-surface grain etching out is largely influenced by the grain energy arising from a high density of defects in the grain structure. **Figure 1** describes, through an E–pH diagram, that aluminum shows passive behavior in a very narrow pH region, at about 4 to 9, which is partially due to the presence of the protective Al_2O_3 film on its surface. The interruptions of the protective oxide film by dissolution occur very quickly wherever the pH is non-neutral – acidic or alkaline – thus causing rapid dissolution. However, the modes of dissolution are pH-dependent; under acidic conditions, aluminum is oxidized to give Al^{3+} ions in solution, whereas under alkaline conditions, the metal dissolves to give aluminate ions (AlO_2^-) in solution.

Therefore, while useful in predicting thermodynamic corrosion susceptibility, Pourbaix diagrams have serious limitations in the practical assessment of aluminum alloys. A crucial issue is the occurrence of localized corrosion across the entire pH range, depending on the applied electrochemical potential, meaning these diagrams cannot reliably be used to determine real corrosion rates. Corrosion rates for individual alloy–electrolyte pairs must be measured independently. Also, practically speaking, none of the available commercial grades of aluminum exhibit immune behavior. Cathodic polarization, wherein such is said to protect an alloy, can

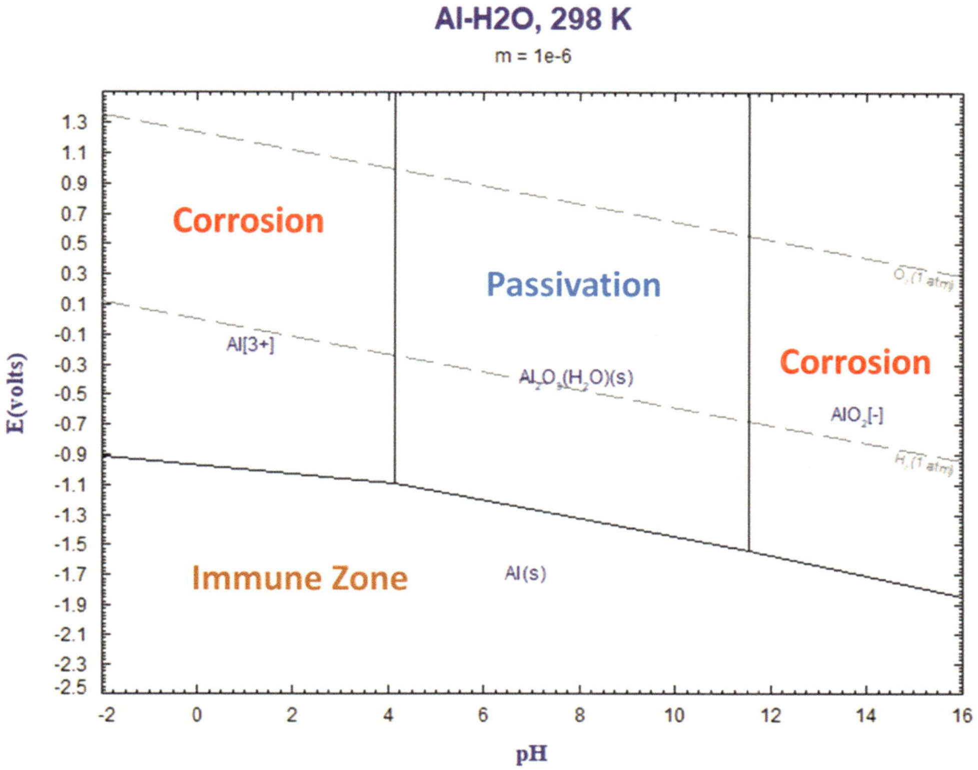

Figure 1.
E–pH curve of aluminum in water at 25°C.

contribute to its deterioration by two main routes: first, the build-up of hydroxyl ions at the surface increases alkaline chemical dissolution of aluminum; and second, aluminum has a strong affinity toward forming hydrides, so hydrogen generated by the cathodic reaction at highly negative potentials may react with the metal to form hydrides that are detrimental to it.

Aluminum corrosion exists in many forms, each having severe practical implications. General dissolution occurs in well-predicted situations in strong acids or alkalis as per Pourbaix diagrams, but with important exceptions: concentrated nitric acid makes aluminum passive through ion transport control of the oxide film, whereas silicate inhibitors may allow the use of alkalis ($pH \leq 11.5$) in chemical handling tanks. Pitting corrosion is the most common and initiates in near-neutral chloride environments, e.g., ships with hulls exposed to seawater or de-iced aircraft with landing gears. Researchers have used XPS and autoradiography to establish chloride adsorption destabilizing passive films, while a point defect theory was proposed to understand the incorporation of Cl^- ions via a vacancy mechanism [8]. Studies of metastable pitting in the form of current oscillations at potentials distinctly below the pitting potential give insights into nano-pit dynamics, with deformed surface layers (100–200 nm thick after rolling or grinding in AA3005 architectural cladding or AA6016 automotive panels) increasing the susceptibility. A rapid failure occurs in bimetallic corrosion whenever aluminum is in contact with nobler metals (copper pipes in marine plumbing, for example, or stainless steel fasteners in offshore platforms), quantified by the ASTM G69 galvanic series. Activators such as Sn/Mg (used in Al-air batteries) or Zn (used in Al–Zn bridge coatings) employ this mechanism intentionally. Crevice corrosion threatens tightly confined spaces (such as aircraft rivet joints or gasket-sealed reactors) based on an oxygen differential aeration mechanism – tightness and depth of the crevice both enhance the severity. Filiform corrosion materializes as thread-shaped filaments beneath coatings (for example, Boeing 737 fuselages or Nordic architectural facades) that progress roughly 100 microns per day away from the defect. Authors have established a direct link to the deformed surface layer and have shown that this corrosion can be alleviated by acid etching (removal of the layer) or by the reduction of Mn content in the alloys. Each mode is instrumental in deciphering the critical interplay among environment, microstructure (e.g., distribution of precipitates), and design for failure, as well as strategies for mitigation, verified using techniques that ranged from electrochemical analyses to synchrotron X-ray investigations.

Alloying elements enhance aluminum's mechanical properties but create electrochemical heterogeneities that are mainly responsible for localized corrosion, such as pitting. Magnesium (Mg) in solid solution slightly retards cathodic processes, but by forming anodic precipitates (Al_3Mg_2/Al_8Mg_5) at the grain boundaries during overaging or overheating of specimens, localized attack is accelerated. Silicon, normally associated with magnesium in 6xxx alloys, forms anodic Mg_2Si particles that preferentially dissolve, as shown in **Figure 2**. Additional silicon increases cathodic kinetics to the detriment of intergranular corrosion. Copper is generally detrimental as it forms highly cathodic intermetallics (e.g., Al_2CuMg, Al_7Cu_2Fe), which efficiently catalyze the reduction of oxygen and therefore accelerate localized dissolution; however, copper in solid solution remains a matter of debate. Zinc increases strength

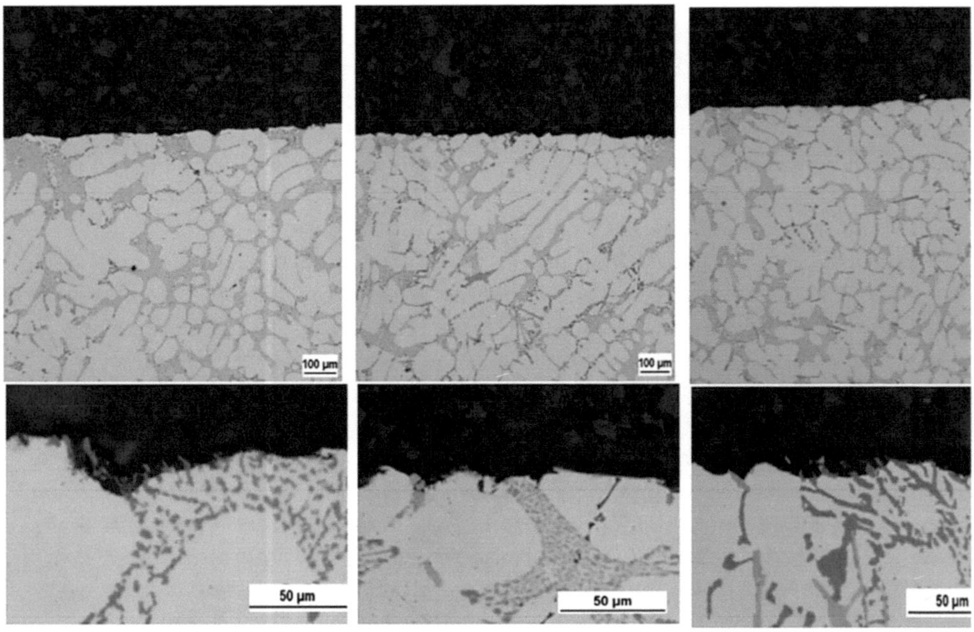

Figure 2.
Optical microscopic images of etched AlSi7Mg0.3 cast alloys with increasing Fe content, showing the corrosion phenomena [10].

by precipitation ($MgZn_2$) in 7xxx alloys, and recently, zinc has been shown to improve SCC resistance in 5xxx alloys by favoring the τ-phase (Al–Mg–Zn) over the susceptible β-phase (Al_3Mg_2). A common form of impurity is iron, which forms cathodic Al_3Fe-type phases that increase local pH and dissolution rates, particularly in the presence of copper. Manganese helps corrosion resistance by altering Fe-rich particles into less noble $Al_6(Mn,Fe)$ constituents, simultaneously diminishing cathodic activity. Lithium decreases density but was also known to cause intergranular corrosion due to anodic Al_3Li; third-generation Al–Li–Cu alloys avoid this issue via processing improvements. Lead-like trace elements disperse passivity through interface segregation with oxides, while tin might somewhat remedy the effects of lead. Small additions (e.g., Zr, Sc, Ti) produce fine, benign Al_3X dispersoids for grain control, thereby increasing corrosion resistance indirectly by suppressing recrystallization. Remarkably, recent studies on sputtered Al-alloys containing refractory metals (W, Mo, Cr) show unusually good passivity, thus guiding the future direction by focusing on microstructural homogeneity with oxide stability. On the whole, the association of anodic or cathodic secondary phases formed by such alloying elements determines corrosion susceptibility [9].

The microstructure of aluminum alloys develops through alloying composition, impurity control, casting procedures, and thermomechanical treatments. Regarding corrosion, the prevalent features of the microstructure are grains and the intermetallic second-phase particles dispersed as constituent particles, dispersoids, or precipitates. These particles have electrochemical properties quite different from those of the matrix alloy; therefore, this phenomenon induces localized galvanic attack termed microgalvanic corrosion. Many investigators have discussed how particular intermetallic particles may influence the corrosion susceptibility of aluminum alloys. Researchers have gathered corrosion potential values for the common intermetallic phases found in aluminum alloys, mostly in chloride-containing solutions. Thereafter, a number of studies have been performed, primarily concerned with the Fe-containing and Cu-containing intermetallics, and hence encompassing syntheses of work on intermetallics in commercial wrought and cast alloys [11]. Identification and characterization of these particles are conducted with X-ray diffraction-based particle extraction, automated electron probe microanalysis, and transmission electron microscopy with X-ray microanalysis. Depending on whether the intermetallic particles are anodic or cathodic with respect to the matrix, two major types of pit morphology are formed. Circumferential pits are one type, comprising a ring of matrix attack that surrounds a more or less intact particle or colony. This galvanically induced corrosion occurs when the noble (cathodic) particle facilitates the dissolution of the more active matrix. The second morphology consists of pits in which the particle is selectively dissolved. These pits are frequently deep and possibly filled with particle remnants and can be interpreted as particle fallout, selective dissolution of electrochemically active particles, or anodic dealloying of Cu-bearing particles with non-faradaic liberation of the copper component. Localized corrosion, as shown in **Figure 3**, still remains a complex phenomenon with continual research aspects. It creates local gradients of pH; increased reduction of oxygen at the cathodic site generates hydroxyl ions, raising the local pH value and in turn affecting the subsequent kinetics of the corrosion propagation and its morphology [12]. It is a question of the precise morphology of the particle-induced pitting that matters in the development of models for the accumulation of damage, requiring a very broad and self-consistent understanding

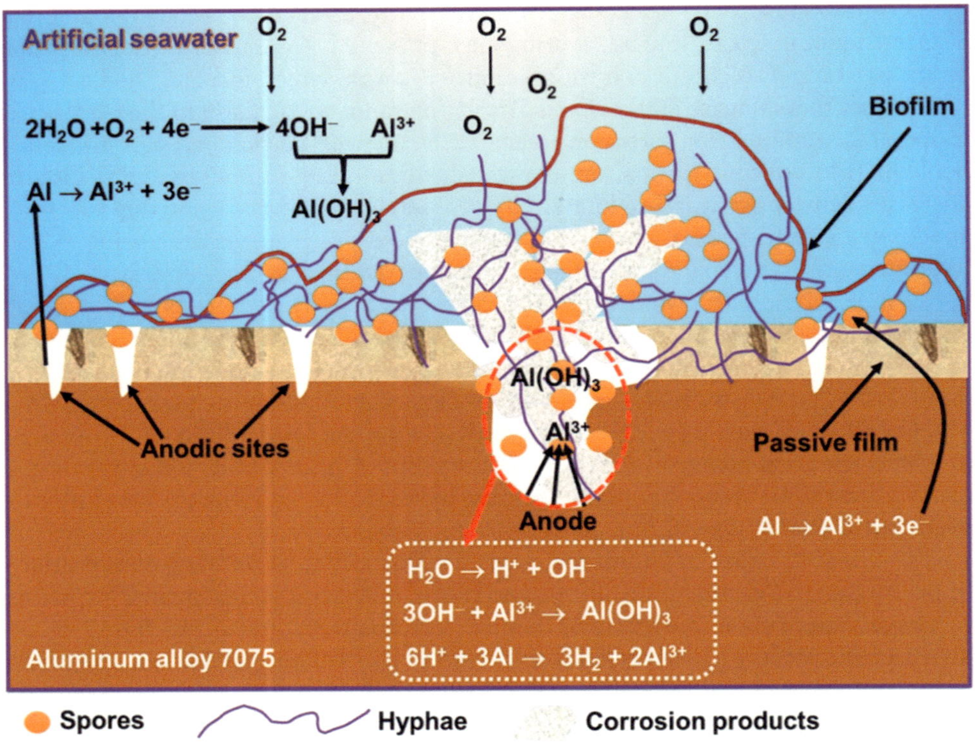

Figure 3.
Mechanism of localized corrosion behavior in aluminum [14].

as the basis. Rigorously characterized constituent particles reveal far more complex behavior in comparison with simple classifications of "noble" or "active." Intermetallics generally play the most important role due to their influence on local corrosion for those noble states that dominate in a pertinent alloy by size or frequency [13]. Thermal treatments induce major changes in local chemistry and electrochemistry and thus alter local corrosion resistance. The time of mechanical working affects grain boundary density (and therefore also grain boundary precipitates) and the distribution and size of constituent particles, which are fragmented and become aligned in the direction of working. Similarly, directionally solidified and segregated microstructures during casting will altogether determine the type, number, and spatial distribution of particles formed within the material.

The corrosion behavior of aluminum alloys is critically dictated by the operational environment and prior processing history. Chlorides present in sea water aggressively promote pitting and crevice-type corrosion by attacking the protective oxide film and promoting microgalvanic coupling with precipitates; galvanic corrosion with metals of higher electrode potential is recognized as a second major concern [15]. Industrial atmospheres comprising SO_2, NO_x, chlorides, and particulates produce general etching and pitting, the intensity of which increases in the presence of moisture-retaining deposits. Underground corrosion (inside soil), on the contrary, varies in scope, but all are driven by low resistivity, moisture, extremes of pH, chlorides/sulfates, MIC, and stray currents. In chemical environments, the amphoteric nature of aluminum makes it susceptible to strong acids (which cause general dissolution, HCl being the worst by virtue of pitting) and strong alkalis (dissolving the oxide); heavy metal ions (Cu^{2+}, Fe^{3+}) precipitate, stimulating severe pitting [16, 17]. High-temperature exposure (>200°C) accelerates oxidation and possibly allows for non-protective scaling, internal oxidation, altered precipitate stability, and hot corrosion via molten salts, with thermal cycling providing further spallation and possibly promoting SCC. More importantly, the processing of alloys (casting, thermomechanical treatment, heat treatment) defines the microstructure: grain structure, type/distribution of intermetallic particles, and residual stress. The microstructure is seen as directly controlling susceptibility to localized attack (pitting and SCC) in such environments. Surface protection (anodizing, conversion coatings, cladding, organic coating) comes next as essential mitigation pathways dependent on the particular environment. Hence, alloy selection, processing route, and protective schemes must all be very carefully optimized for the particular service conditions.

3. Corrosion protection methodologies

The corrosion of aluminum and its alloys in aggressive aqueous environments is an issue of primary concern in industries such as aerospace, automotive, and construction. The major aspects of its prevention mostly emphasize or are based on soluble chemical inhibitors, defined as substances soluble in water that inhibit the electrochemical reactions constituting the activities of the corrosion cell. Yet, this classification still holds in the fundamental sense of an inhibitor acting to suppress the anodic (metal dissolution) or the cathodic (usually oxygen or hydrogen reduction) reactions, or both (mixed inhibitors). In recent years, however, views on inhibition mechanisms have been enormously broadened, and alternatives to the traditionally used toxic substances have been developmentally pursued, preferably on an environmentally compatible basis [18]. Anodic inhibitors act quite basically by either promoting oxide film formation or repairs so that the pitting potential is raised and pit initiation is dispersal-determined through electrochemical or exposure testing. However, in modern times, it has been shown that the major relevance of local chemistry and surface heterogeneity is well understood, and the activation of pitting by pre-existing flaws or inclusions is still pertinent. More importantly, established pits propagate with little hindrance, as these inhibitors generally do not promote repassivation within active pits. Cathodic inhibitors operate mainly by blocking the cathodic half-reaction (e.g., oxygen reduction) and thereby necessitating a simultaneous decrease in the anodic dissolution rate and lowering of the free corrosion potential. In fact, from recent electrochemical studies, it was found that the best cathodic inhibitors cause the corrosion potential to shift far below the pitting or repassivation potential of the alloy. At the nanoscale, they have proved their key advantage – very high effectiveness at very low concentrations – as demonstrated using an SECM (scanning electrochemical microscope). The standard that once fit this category was hexavalent chromium (chromate), a well-known effective inhibitor of oxygen reduction at micromolar concentrations in chloride solutions. However, the environmental concerns raised, the adoption of RoHS regulations, and occupational health issues related to chromate's carcinogenicity have led to intense research over the last two decades into finding and optimizing suitable alternatives.

Some major achievements attained in this research are inorganic alternatives involving rare-earth cations (Ce^{3+}, La^{3+}) [19]. They are commonly administered in the form of cerium carboxylates or otherwise encapsulated within nanostructures for controlled release and are promising candidates as cathodic inhibitors; they operate by generating hydroxides/oxides at sites of cathodic activity. Molybdate and tungstate ions serve as mixed-type inhibitors, and recent studies have focused mainly on synergistic mixtures with silicates or organic compounds to boost performance under neutral/alkaline environments. Vanadates and permanganates are also of high research interest, although their environmental aspects raise concerns. Organic inhibitors form a large and ever-increasing arena, including both specially synthetic molecules (phosphonates, e.g., HEDP; sulfonates; azoles such as benzotriazole/BTA and its derivatives; imidazoles; thiols; and various amines) and well-studied natural extracts (rich in polyphenols, alkaloids, or flavonoids) extracted from different plants (leaves, roots, or even fruit peels). Research concerning adsorption mechanisms (physisorption, chemisorption via heteroatoms like N, O, S, P), their structure-activity relationships, and their ability to form protective films occupies the main research theme. Synergistic blends are increasingly gaining popularity, where organic molecules are combined with inorganic ions (e.g., cerium-BTA, molybdate-phosphonate), potentially engaging several inhibitive pathways to provide enhanced inhibition and reduce inhibitor dosage. Novel methods of delivery are paramount to their effectiveness and longevity. Traditionally, inhibitors are directly added into the environment; nowadays, increasing attention is given to incorporating inhibitors into protective coating systems. Traditionally, this is done via pigment-based approaches with sparingly soluble salts (e.g., strontium phosphosilicate as a replacement for strontium chromate) or by ion-exchange pigments (e.g., $Mg_2Al^{-}NO_3^{-}$ LDHs exchanging nitrate for corrosive chlorides). More advanced research is exploring "smart" release mechanisms based on nanocontainers (mesoporous silica, halloysite nanotubes, polyelectrolyte capsules, LDHs) or nanocomposite coatings that release inhibitors (organic or inorganic) only when they are summoned by particular corrosion triggers like pH changes or chloride ingress, minimizing leaching and extending service life [20]. Surface pre-treatments with inhibitor-loaded sol-gel coatings and direct adsorption from evaporable solvents also figure as important application methods for inhibitors. To be brief, contemporary research focuses on environmentally acceptable and high-speed inhibitors (both inorganic and organic agents), synergism, and advanced responsive delivery systems to provide effective and sustainable corrosion protection, while the principles of aluminum corrosion inhibition have remained untouched throughout.

4. Challenges and future scope

The prediction and mitigation of localized corrosion owing to alloy microstructure and intermetallics require accurate analysis, which can help in preventing unexpected failure in service, traditionalist designs, and gaps between lab-oriented data. Localized pitting and SCC, deemed the worst forms of failures, are intrinsically based on the local microstructural heterogeneity. Intermetallic particles and precipitates act as tiny micro-galvanic cells, causing the aggressive localized attack that is often quickly set off, which the usual barrier coating and bulk inhibitor fail to avert. This is worsened when there is mechanical stress to further speed up the pit-to-crack transition; at the same time, SCC under cyclic loads severely undermines any heavy application of protective coatings and is nearly impossible to predict. Researchers have proposed improved active protection systems such as smart coatings with leachable inhibitors from Li salts, azoles, or nanocontainers (MOFs, halloysite nanotubes, etc.). Yet, all of them come with serious limitations, such as uncontrolled leaching kinetics, which limit their longevity, while many of the active compounds themselves raise serious environmental and regulatory concerns due to their ecotoxicity and long-term fate. Integrating nanocarriers into coating matrices is still a poorly understood subject, usually compromising the mechanical properties of the matrices and providing pathways for moisture ingress.

Another bottleneck for the progression of the field is the separation between laboratory research and industrial realization. Many of the breakthrough coatings demonstrated in laboratories depend on complex synthesis methods, which are unable to be scaled and lack mechanical robustness. Their performance is most commonly measured using standardized accelerated tests, such as Neutral Salt Spray, which are known not to correlate with real-time service environments and have only led to either over-engineered designs or unexpected field failures. The corrective approach would necessitate a multidimensional integrated research agenda, using high-resolution operando characterization techniques to create a direct link between microstructure and electrochemical activity, and creating multi-scale physics-based models, which couple quantum-mechanical calculations of reactivity with continuum models for transport and damage. At the same time, the community must focus on the development of environmentally friendly "green" inhibitors and on creating standardized datasets that are ready for machine learning.

5. Conclusions

The corrosion phenomena of aluminum alloys depend primarily on alloy chemistry and microstructure, which control mechanical properties as well as corrosion susceptibility. Though much work has gone into initiation mechanisms and various ways of protection, some key propagation elements such as intergranular corrosion, exfoliation, SCC, and hydrogen embrittlement need to be considered for structural applications and investigated further. In advance, corrosion protection strategies have achieved great improvement, and the advent of state-of-the-art techniques in electrochemistry and surface characterization has aided immensely in the search for sophisticated solutions, particularly in chromate inhibitor replacement and smart coating technology, which are absolutely necessary so that these innovations can meet the current limitations in environmental as well as performance standards. In the future, one considerable challenge would appear to be the realization of ultra-high strength (>1,000 MPa) aluminum alloys. The usual precipitation hardening method might not be able to give the desired results; therefore, other methods such as severe plastic deformation or rapid quenching to form amorphous alloys could be explored. Amorphous Al-based systems (alloyed with Zr or Ti or Nb or rare earths) have tremendous strength potential (up to 1,500 MPa), though their application is practically hampered by some major unresolved problems, as the thermal stability offered to the amorphous phases is quite limited, the fabrication processes needed for large-scale components are undefined, and, most importantly, there is a great lack of understanding about their corrosion mechanisms and long-term sustainability in a service environment. Future development is hence going to depend upon alloy design for corrosion resistance, the evolution of sound environmentally compatible protection systems, and, of course, the scaling up and stabilizing of the next generation of ultra-high strength aluminum materials.

Conflict of Interest

The authors declare no conflict of interest.

Author details

Amlan Das[1] and Shashanka Rajendrachari[2*]

1 Mechatronics Department, G. H. Patel College of Engineering and Technology, The Charutar Vidya Mandal (CVM) University, Anand, Gujarat, India

2 Department of Chemistry, Global Academy of Technology, Bangalore, Karnataka, India

*Address all correspondence to: shashankaic@gmail.com

References

[1] Jeeva N, Thirunavukkarasu K, Xavier JR. Influence of multifunctional graphene oxide and silanized vanadium nitride in polyurethane coatings for the protection of aluminium alloy in aerospace industries. Diamond and Related Materials. 2024;**142**:110792. DOI: 10.1016/j.diamond.2024.110792

[2] Cao H, Ma X, Wang T, Jia W, Gong B, Guo L, Xue P, Zhu C, Tian X, Min Y, Xu Q. Graphene-like 2D Ti3C2Tx MXene/dodecyltrimethoxysilane coating for corrosion protection on aluminum alloy by designing multilayer structure. Diamond and Related Materials. 2024;**141**:110696. DOI: 10.1016/j.diamond.2023.110696

[3] Zhang S, Ma Y, Zhao S, Dai W, Liu B. Enhanced coating system corrosion protection of 2A14 Al alloy by modified cerium seal PEO pretreatment. Corrosion Science. 2024;**235**:112203. DOI: 10.1016/j.corsci.2024.112203

[4] Ma X, Wang T, Gong B, Cao H. Fast and low-cost deposition strategy for constructing amorphous carbon layer toward corrosion protection on aluminum alloy bipolar plates in proton exchange membrane fuel cell environments. Journal of Power Sources. 2024;**623**:235479. DOI: 10.1016/j.jpowsour.2024.235479

[5] Wang Y, Zhou J, Zhou B, Li Y, He K, Wei Y. Electrodeposition preparation of ZnAlCe-LDH film for corrosion protection of 6061 Al alloy. Materials Letters. 2024;**359**:135965. DOI: 10.1016/j.matlet.2024.135965

[6] Lavanya M, Ghosal J, Rao P. A comprehensive review of corrosion inhibition of aluminium alloys by green inhibitors. Canadian Metallurgical Quarterly. 2024;**63**:119–129. DOI: 10.1080/00084433.2023.2190198

[7] Li Z, Visser P, Hughes AE, Homborg A, Gonzalez-Garcia Y, Mol A. Review of the state of art of Li-based inhibitors and coating technology for the corrosion protection of aluminium alloys. Surface and Coatings Technology. 2024;**478**:130441. DOI: 10.1016/j.surfcoat.2024.130441

[8] Okayasu M, Takeuchi S, Shiraishi T. Corrosion and mechanical properties of cast aluminium alloys. International Journal of Cast Metals Research. 2013;**26**:319–329. DOI: 10.1179/1743133613Y.0000000065

[9] Reboul MC, Baroux B. Metallurgical aspects of corrosion resistance of aluminium alloys. Materials and Corrosion. 2011;**62**:215–233. DOI: 10.1002/maco.201005650

[10] Kuchariková L, Liptáková T, Tillová E, Kajánek D, Schmidová E. Role of chemical composition in corrosion of aluminum alloys. Metals (Basel). 2018;**8**:581. DOI: 10.3390/met8080581

[11] Donatus U, Thompson GE, Omotoyinbo JA, Alaneme KK, Aribo S, Agbabiaka OG. Corrosion pathways in aluminium alloys. Transactions of Nonferrous Metals Society of China. 2017;**27**:55–62. DOI: 10.1016/S1003-6326(17)60006-2

[12] Khedr MGA, Lashien AMS. The role of metal cations in the corrosion and corrosion inhibition of aluminium in aqueous solutions. Corrosion Science. 1992;**33**:137–151. DOI: 10.1016/0010-938X(92)90023-V

[13] Ambat R, Davenport AJ, Scamans GM, Afseth A. Effect of iron-containing intermetallic particles on the corrosion behaviour of aluminium. Corrosion Science. 2006;**48**:3455–3471. DOI: 10.1016/j.corsci.2006.01.005

[14] Zhang Y, He J, Zheng L, Jin Z, Liu H, Liu L, Gao Z, Meng G, Liu H, Liu H. Corrosion of aluminum alloy 7075 induced by marine Aspergillus terreus with continued organic carbon starvation. Npj Materials Degradation. 2022;**6**:27. DOI: 10.1038/s41529-022-00236-2

[15] Šekularac G, Milošev I. Corrosion of aluminium alloy AlSi7Mg0.3 in artificial sea water with added sodium sulphide. Corrosion Science. 2018;**144**:54–73. DOI: 10.1016/j.corsci.2018.08.038

[16] Syed S. Influence of the environment on atmospheric corrosion of aluminium, Corrosion Engineering. Science and Technology. 2010;**45**:282–287. DOI: 10.1179/147842208X373182

[17] Deepa P, Padmalatha R. Corrosion behaviour of 6063 aluminium alloy in acidic and in alkaline media. Arabian Journal of Chemistry. 2017;**10**: S2234–S2244. DOI: 10.1016/j.arabjc.2013.07.059

[18] Carreira AF, Pereira AM, Vaz EP, Cabral AM, Ghidini T, Pigliaru L, Rohr T. Alternative corrosion protection pretreatments for aluminum alloys. Journal of Coatings Technology and Research. 2017;**14**:879–892. DOI: 10.1007/s11998-017-9922-9

[19] Allachi H, Chaouket F, Draoui K. Protection against corrosion in marine environments of AA6060 aluminium alloy by cerium chlorides. Journal of Alloys and Compounds. 2010;**491**: 223–229. DOI: 10.1016/j.jallcom.2009.11.042

[20] Andreeva DV, Skorb EV. Multi-layer smart coatings for corrosion protection of aluminium alloys and steel. In: Abdel Salam, Hamdy Makhlouf, editors. Handbook of Smart Coatings for Materials Protection. Cambridge, UK: Woodhead Publishing; 2014. p. 307–327. DOI: 10.1533/9780857096883.2.307